Wearable Technologies

Wearable Technologies

Special Issue Editors

Nicola Carbonaro
Alessandro Tognetti

MDPI • Basel • Beijing • Wuhan • Barcelona • Belgrade

MDPI

Special Issue Editors

Nicola Carbonaro
University of Pisa
Italy

Alessandro Tognetti
University of Pisa
Italy

Editorial Office
MDPI
St. Alban-Anlage 66
4052 Basel, Switzerland

This is a reprint of articles from the Special Issue published online in the open access journal *Technologies* (ISSN 2227-7080) from 2017 to 2018 (available at: https://www.mdpi.com/journal/technologies/special_issues/wearable_technologies)

For citation purposes, cite each article independently as indicated on the article page online and as indicated below:

LastName, A.A.; LastName, B.B.; LastName, C.C. Article Title. *Journal Name* **Year**, *Article Number, Page Range.*

ISBN 978-3-03897-513-7 (Pbk)
ISBN 978-3-03897-514-4 (PDF)

Contents

About the Special Issue Editors

Nicola Carbonaro, PhD, is Assistant Professor at the Information Engineering Department and the Research Center "E. Piaggio" of the University of Pisa. He graduated in Electronic Engineering at the University of Pisa in 2004. In 2010, he earned a PhD in Information Engineering from the University of Pisa working on the development of wearable systems for human activity classification. In 2009, he spent six months as a visiting researcher at the "Neural Control of Movement" Laboratory of Arizona State University. His research is mainly focused on hardware and software development for wearable sensing technology for physiological and behavioral human monitoring for biomedical applications. Since 2014, he is the chair of Biosensors, as part of the Biomedical Engineering Degree of the University of Cagliari. Dr. Carbonaro has collaborated on different research projects both at a National and European level and he has published several papers, contributions to international conferences, and book chapters.

Alessandro Tognetti, PhD, is Assistant Professor at the Information Engineering Department and the Research Center "E. Piaggio" of the University of Pisa. He graduated in Electronic Engineering and he received his PhD degree in Robotics, Automation, and Bioengineering from the University of Pisa in 2005. He is the chair of Biosensors and teaches Bioelectrical Phenomena on the Biomedical Engineering Degree, under the School of Engineering, at the University of Pisa. His competences and skills range from sensor build up—starting from the physical principle and the enabling technology—system integration, and high-level interpretation/fusion to the development of ICT-supported applications in e-health, rehabilitation, robotics, and human/machine interaction. His research activities have resulted in more than 100 international scientific publications. He carried out most of his research in the frame of European and International projects (20 projects of which 12 were European), in which he participated as the team leader. Among these projects he was the work-package leader of the EU IP project ProeTex, leading a group of 12 partners, developing sensors and biosensors for emergency personnel monitoring.

technologies

MDPI

Editorial

Special Issue on "Wearable Technologies"

Alessandro Tognetti [1,2,*] and Nicola Carbonaro [1,2]

[1] Dipartimento Ingegneria dell'Informazione, Università di Pisa, University of Pisa, Largo Lucio Lazzarino 1, 56122 Pisa, Italy; nicola.carbonaro@unipi.it
[2] Centro di Ricerca "E.Piaggio", University of Pisa, Largo Lucio Lazzarino 1, 56122 Pisa, Italy
* Correspondence: alessandro.tognetti@unipi.it

Received: 5 November 2018; Accepted: 7 November 2018; Published: 8 November 2018

Wearable technology will revolutionize our lives in the years to come. The current trend is to augment ordinary body-worn objects—e.g., watches, glasses, bracelets, and clothing—with advanced information and communication technologies (ICT) such as sensors, electronics, software, connectivity and power sources. These wearable devices can monitor and assist the user in the management of his/her daily life with applications that may range from activity tracking, sport and wellness, mobile games, environmental monitoring, up to eHealth.

The present Special Issue reports the recent advances in the multidisciplinary field of wearable technologies and the important gaps that still remain in order to obtain a massive diffusion.

In the frame of wearable technologies, this Special Issue of *Technologies* includes a total of 10 papers, including one review paper and nine research articles. Articles in this Special Issue address topics that include: wearable sensing and bio-sensing technologies, smart textiles, smart materials, wearable microsystems, low-power and embedded circuits for data acquisition and processing and data transmission.

The first feature paper from Münzenrieder et al. [1] focusses on advanced technologies to push forward the smart textile field. In the presented research, the authors benchmarked different fabrication techniques and multiple fibers made from polymers, cotton, metal and glass exhibiting diameters down to 125 μm, to obtain fully functional transistor fibres. In particular, by exploiting the most promising fabrication approach, they were able to integrate a commercial nylon fiber functionalized with InGaZnO TFTs into a woven textile. The second feature paper is from Santos et al. [2] and it presents a methodology for movement recognition in hand-assisted laparoscopic surgery using a textile-based sensing glove. The aim is to recognize the commands given by the surgeon's hand inside the patient's abdominal cavity in order to guide a collaborative robot. The glove, which incorporates piezoresistive sensors, continuously captures the degree of flexion of the surgeon's fingers. These data are analyzed throughout the surgical operation using an algorithm that detects and recognizes some defined movements as commands for the collaborative robot. The results obtained with 10 different volunteers showed a high degree of precision and recall.

Wearable technologies are fundamental building blocks for the Virtual Reality (VR) and Augmented Reality (AR) fields as underlined in the next two contributions. The work from Cutolo et al. [3] reports an innovative hybrid video-optical see-through Head Mounted Display (HMD). The geometry of the HMD explicitly violates the rigorous conditions of orthostereoscopy. For properly recovering natural stereo fusion of the scene within the personal space in a region around a predefined distance from the observer, the authors partially resolved the eye-camera parallax by warping the camera images through a perspective preserving homography that accounts for the geometry of the video see-through HMD and refers to such distance. The results obtained showed that the quasi-orthoscopic setting of the HMD; together with the perspective preserving image warping; allow the recovering of a correct perception of the relative depths. The paper of Maereg et al. [4] presents a low cost, wearable six Degree of Freedom (6-DOF) hand pose tracking system for Virtual

Reality applications. The wearable system is designed for use with an integrated hand exoskeleton system for kinesthetic haptic feedback. The tracking system consists of an Infrared (IR) based optical tracker with low cost mono-camera and inertial and magnetic measurement unit. Six DOF hand tracking outputs filtered and synchronized on LabVIEW software are then sent to the Unity Virtual environment via User Datagram Protocol (UDP) stream. Experimental results show that this low cost and compact system has a performance that makes it fully suitable for VR applications.

The next four contributions deal with applications of wearable technologies in the eHealth sector. The paper from Signorini et al. [5] describes a methodology for prenatal monitoring of fetal heart rate (FHR). As underlined by the authors, a wearable system able to continuously monitor FHR would be a noticeable step towards a personalized and remote pregnancy care. The wearable system presented employs textile electrodes and miniaturized electronics integrated in smart platform enabled by mobile devices. The system has been tested on a limited set of pregnant women whose fetal electrocardiogram recordings were acquired and classified, yielding an overall score for both accuracy and sensitivity over 90%. This novel approach can open a new perspective on the continuous monitoring of fetus development by enhancing the performance of regular examinations, making treatments really personalized, and reducing hospitalization or ambulatory visits. Another branch of eHealth is the monitoring of elderly people to early detect symptoms related to possible health treats (e.g., frailty, falls, dementia, etc.). In this context, Genovese et al. in [6] reports the sensor description and the preliminary testing of a an integrated fall detection and prevention ICT service for elderly people based on wearable smart sensors. Falls are one of the most common causes of accidental injury: approximately, 37.3 million falls requiring medical intervention occur each year. Fall-related injuries may cause disabilities, and in some extreme cases, premature death among older adults, which has a significant impact on health and social care services. The fall detector is intended to be worn at the waist level for use during activities of daily living; a dedicated logger is intended for the quantitative assessment of tested individuals during the execution of clinical tests. Both devices provide their service in conjunction with an Android mobile device. The work from Bock et al. [7] investigates on the reliability of consumer-grade physical activity monitors (CPAMs). The study is performed on thirty subjects that wore different activity monitors (a total of eight monitors are employed). The wearable devices were tested in the lab and in free-living setting. The results shown that all activity monitors yield reliable estimations of physical activity. However, all CPAMs tested provided reliable estimations of physical activity within the laboratory but appeared less reliable in a free-living setting. Finally, the eHealth section of this special issue includes the review paper from Sharma et al. [8]. This review paper focusses on a hot topic of the biomedical technology: cuffless and continuous monitoring of blood pressure (BP). As underlined by the authors, in the recent years, the indirect approach to obtain BP values has been intensively investigated, where BP is mathematically derived through the "Time Delay" in propagation of pressure waves in the vascular system, obtaining cuffless and continuous BP monitoring. The review highlights recent efforts in developing these next-generation blood pressure monitoring devices and compares various mathematical models. The unmet challenges and further developments that are crucial to develop cuffless BP devices are also discussed.

The paper from Ben Arbia et al. [9] investigates on wearable wireless networks (WWNs) as innovative ways to connect humans and/or objects anywhere, anytime, within an infinite variety of applications. In particular, the authors performed experiments on a real testbed to investigate the connectivity behavior on two wireless communication levels: on-body and body-to-body.

Flexible and stretchable materials and sensing substrates are a relevant topic in the wearable technology field, with potential of opening new applications in human bio-monitoring and human machine interaction. In this context, the work from Russo et Al [10] presents a stretchable tactile sensor based on electrical impedance tomography (EIT), an imaging method that can be applied over stretchable conductive-fabric materials to realize soft and wearable pressure sensors through current injections and voltage measurements at electrodes placed at the boundary of a conductive medium.

The articles published in this Special Issue present detailed views of some of the most important topics about wearable technologies underlining potential applications for the health and AR/VR sectors. Integration of sensors into flexible/stretchable substrates, such as textiles, will further increase the widespread diffusion of wearable technologies.

Acknowledgments: The Guest Editors would like to thank all the authors for their invaluable contributions and the anonymous reviewers for their fundamental suggestions and comments.

Conflicts of Interest: The authors declare no conflict of interest.

References

1. Münzenrieder, N.; Vogt, C.; Petti, L.; Salvatore, G.A.; Cantarella, G.; Büthe, L.; Tröster, G. Oxide Thin-Film Transistors on Fibers for Smart Textiles. *Technologies* **2017**, *5*, 31. [CrossRef]
2. Santos, L.; Carbonaro, N.; Tognetti, A.; González, J.L.; de la Fuente, E.; Fraile, J.C.; Pérez-Turiel, J. Dynamic Gesture Recognition Using a Smart Glove in Hand-Assisted Laparoscopic Surgery. *Technologies* **2018**, *6*, 8. [CrossRef]
3. Cutolo, F.; Fontana, U.; Ferrari, V. Perspective Preserving Solution for Quasi-Orthoscopic Video See-Through HMDs. *Technologies* **2018**, *6*, 9. [CrossRef]
4. Maereg, A.T.; Secco, E.L.; Agidew, T.F.; Reid, D.; Nagar, A.K. A Low-Cost, Wearable Opto-Inertial 6-DOF Hand Pose Tracking System for VR. *Technologies* **2017**, *5*, 49. [CrossRef]
5. Signorini, M.G.; Lanzola, G.; Torti, E.; Fanelli, A.; Magenes, G. Antepartum Fetal Monitoring through a Wearable System and a Mobile Application. *Technologies* **2018**, *6*, 44. [CrossRef]
6. Genovese, V.; Mannini, A.; Guaitolini, M.; Sabatini, A.M. Wearable Inertial Sensing for ICT Management of Fall Detection, Fall Prevention, and Assessment in Elderly. *Technologies* **2018**, *6*, 91. [CrossRef]
7. Bock, J.M.; Kaminsky, L.A.; Harber, M.P.; Montoye, A.H.K. Determining the Reliability of Several Consumer-Based Physical Activity Monitors. *Technologies* **2017**, *5*, 47. [CrossRef]
8. Sharma, M.; Barbosa, K.; Ho, V.; Griggs, D.; Ghirmai, T.; Krishnan, S.K.; Hsiai, T.K.; Chiao, J.-C.; Cao, H. Cuff-Less and Continuous Blood Pressure Monitoring: A Methodological Review. *Technologies* **2017**, *5*, 21. [CrossRef]
9. Arbia, D.B.; Alam, M.M.; Moullec, Y.L.; Hamida, E.B. Communication Challenges in on-Body and Body-to-Body Wearable Wireless Networks—A Connectivity Perspective. *Technologies* **2017**, *5*, 43. [CrossRef]
10. Russo, S.; Nefti-Meziani, S.; Carbonaro, N.; Tognetti, A. Development of a High-Speed Current Injection and Voltage Measurement System for Electrical Impedance Tomography-Based Stretchable Sensors. *Technologies* **2017**, *5*, 48. [CrossRef]

technologies

MDPI

Article

Oxide Thin-Film Transistors on Fibers for Smart Textiles

Niko Münzenrieder [1,2,*], **Christian Vogt** [2], **Luisa Petti** [2], **Giovanni A. Salvatore** [2], **Giuseppe Cantarella** [2], **Lars Büthe** [2] **and Gerhard Tröster** [2]

1   Sensor Technology Research Centre, University of Sussex, Falmer BN1 9QT, UK
2   Electronics Laboratory, Swiss Federal Institute of Technology, Zürich 8092, Switzerland;
    christian.vogt@ife.ee.ethz.ch (C.V.); luisa.petti@ife.ee.ethz.ch (L.P.);
    giovanni.salvatore@ife.ee.ethz.ch (G.A.S.); giuseppe.cantarella@ife.ee.ethz.ch (G.C.);
    lars.buethe@ife.ee.ethz.ch (L.B.); troester@ife.ee.ethz.ch (G.T.)
*   Correspondence: n.s.munzenrieder@sussex.ac.uk; Tel.: +44-127-387-2631

Received: 30 April 2017; Accepted: 29 May 2017; Published: 2 June 2017

Abstract: Smart textiles promise to have a significant impact on future wearable devices. Among the different approaches to combine electronic functionality and fabrics, the fabrication of active fibers results in the most unobtrusive integration and optimal compatibility between electronics and textile manufacturing equipment. The fabrication of electronic devices, in particular transistors on heavily curved, temperature sensitive, and rough textiles fibers is not easily achievable using standard clean room technologies. Hence, we evaluated different fabrication techniques and multiple fibers made from polymers, cotton, metal and glass exhibiting diameters down to 125 µm. The benchmarked techniques include the direct fabrication of thin-film structures using a low temperature shadow mask process, and the transfer of thin-film transistors (TFTs) fabricated on a thin ($\approx$1 µm) flexible polymer membrane. Both approaches enable the fabrication of working devices, in particular the transfer method results in fully functional transistor fibers, with an on-off current ratio $>10^7$, a threshold voltage of $\approx$0.8 V, and a field effect mobility exceeding $7\,\mathrm{cm^2\,V^{-1}\,s^{-1}}$. Finally, the most promising fabrication approach is used to integrate a commercial nylon fiber functionalized with InGaZnO TFTs into a woven textile.

Keywords: field-effect transistors; thin-film technology; InGaZnO; oxide semiconductors; smart textiles

1. Introduction

Electronic or smart textiles (e-textiles) promise to have a significant impact in areas such as wearable computing or large-area electronics [1]. Potential areas of application include healthcare, sports, or support of high risk professionals, e.g., firefighters [2–4]. Here, the vision is of an e-textile consisting of a fabric that preserves all the properties of textile fibers, such as comformability, washability, softness or stretchability, and combines them with electronic functionality. The aforementioned electronic functionality often refers to different sensors e.g., for strain, posture, temperature or other physiological signals [5,6] but also includes the associated conditioning circuits, power supply, and signal processing or transmission electronics [7–9]. So far, the spectrum of e-textiles ranges from conventional electronics attached to textiles [10] to electronic components build from active textile yarns [11,12]. The first approach, usually realized by integrating rigid off-the-shelf electrical devices and circuit boards, drastically influences the mechanical properties of the textile, while, the second one in general only provides limited electronic complexity and hence limited electronic performance [13]. An alternative approach is the integration of flexible electronics into a woven textile. Here, the use of flexible plastic stripes as carriers for thin-film devices and standard silicon chips, represents a good compromise between the mechanical and electrical properties of the final textile device [14]. Additionally, the integration of electronic fibers and conductive yarns in the weft and warp direction of a woven fabric also enables the fabrication of more complex systems inside a textile. Nevertheless, the integration of flexible stripes causes another fabric specific problem which is in particular important concerning the mass production of electronic textiles: Non-circular fibers such as planar plastic stripes are not compatible with standard weaving equipment, and are sensitive to twisting which calls for modified knitting or embroidery machines [15].

The solution to this problem is the fabrication of mechanically flexible active electronic devices directly on circular fibers. Since the fabrication of electronic devices on fibers, compatible with the demands of the textile industry, is challenging only few associated reports including a temperature sensor on a nylon yarn have been published [16]. In this context, the fabrication related challenges arise from the required flexibility, and the chemical and physical proprieties of the available yarns. Additionally, yarns usable for the fabrication of textiles exhibit diameters significantly below 1 mm, which results in a highly curved surface. These challenges can be addressed by new developments in the area of flexible electronics. In particular the use of oxide semiconductors, such as amorphous InGaZnO (IGZO) [17–19], promises to realize high performance active electronic devices on a variety of substrates. Here, we evaluated how IGZO thin-film transistors (TFTs), representing the most important and basic building block of all electronic systems, can be fabricated on a variety of different yarns. It is shown that high performance TFTs, on glass fibers with a radius of 62.5 µm and on polymer fibers with a radius of 125 µm, are fully functional and can be integrated into textiles for wearable or industrial applications.

2. Fabrication of TFTs on Fibers

In contrast to conventional substrates used for the fabrication of electronic thin-film devices, such as semiconductor wafers, glass plates or plastic foils, the mechanical and geometrical properties of fibers and yarns are less beneficial. Hence, the successful fabrication of transistors requires a modification of the fabrication process and a proper selection of suitable yarns or fibers. Here, technologies developed for the fabrication of flexible and stretchable electronics are adapted.

2.1. Micro Processing on Yarns and Fibers

We evaluated a range of possible substrate fibers. As shown in Figure 1a, these included steel and cotton yarns, nylon fibers with different diameters, glass fibers, and thin insulated metal Cu (magnet) wire. All materials have certain advantages and disadvantages concerning the fabrication of smart textiles. The most important parameters for the fabrication of TFTs and electronic textiles are:

- Chemical properties: The chemical stability of the fiber material is a key aspect since the fibers have to resist the etchants and solvents used during the fabrication process. In this respect the metal and glass fibers exhibit the most beneficial properties.
- Temperature resistance: Similar to the chemical properties, the melting or glass transition temperature of the evaluated materials can significantly limit the choice of usable deposition technologies. While the maximum temperature of cotton and nylon is in the range of 200 °C, the glass fiber can be processed at temperatures above 1000 °C.
- Fiber surface: Thin-film devices are made from active layers with thickness in the nanometer range, hence the surface of the fibers has to be as flat as possible. While the steel and cotton yarns do not exhibit a continuous surface, also the surface roughness of the other fibers varies strongly. The rms value of the employed glass fibers is <10 nm, but the corresponding values for nylon and the insulated Cu wire reach values of 10 µm and 1 µm, respectively.
- Conductivity: Non-conductive fibers (glass, cotton, nylon) have the advantage that no additional insulation layer is needed, and all electronic devices on their surface are decoupled from each other. Metallic substrate fibers at the same time, could simplify the device structure

by providing electronic functionality themselves. Here an interesting option could be the use the insulated Cu wire as substrate fiber, gate contact and gate insulator simultaneously.

- Textile properties: Unobtrusive smart textiles call for electronic fibers which are soft, bendable, and with dimensions comparable to the textile yarns of the fabric. In this respect cotton but also steel yarns have beneficial properties. Similarly, polymer fibers such as nylon are common. Anyway, the diameter of the nylon fibers should not be too large ($\lesssim$750 µm [20]). Furthermore, thin Cu wires are bendable and can be imperceptible when integrated into a textile [21]. Glass fibers on the other hand exhibit a small diameter, but their minimum bending radius is limited to $\approx$5 cm.

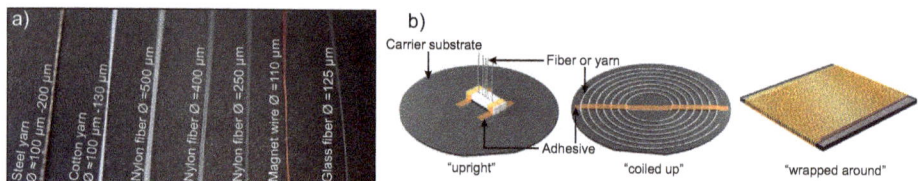

Figure 1. Thin-film technology on fibers: (**a**) Photograph of the fibers and yarns evaluated as substrate fibers for the fabrication of thin-film devices. (**b**) Different approaches to load flexible fibers into standard semiconductor manufacturing equipment.

In total it can be concluded that the continuous cylindrical shape, the wide availability, the variable diameter, the mechanical flexibility, and its use in commercial textiles makes nylon the most suitable choice for the fabrication of electronic fibers. At the same time, the high surface roughness of commercial nylon fibers remains an issue.

Another issue which has to be considered is the extreme form factor (relation between diameter and length) of all kinds of fibers. First it has to be mentioned that the most effective solution for the fabrication of long functionalized fibers, desirable for the fabrication of textiles, would be roll-to-roll fabrication [22]. Specialized equipment to continuously coat fibers has been developed using for example sputtering techniques [23]. Loading a fiber into a commercial semiconductor processing tool, and structuring the deposited layers, in general requires the use of a carrier substrate to provide mechanical support and to simplify the handling of the fiber during the fabrication process. Here we considered three basic possibilities, illustrated in Figure 1b, to ensure comparability between the substrate fibers and the processing equipment. Mounting short fibers upright on the carrier enables a 360° coating of the fibers, but also limits their length which is contradictory to their use in a textile. Coiling up the fiber on the surface of a carrier allows processing of longer fibres, the disadvantages are that only one halve of the fiber surface (top side) is coated, and that there is mechanical strain induced all along the fiber. Finally wrapping the fiber around a carrier substrate can be used for very long fibers (a 3 inch carrier substrate in combination with a 250 µm fiber and a 50 % fill factor results in a max fiber length of $\approx$20 m). The disadvantages are that again only one halve of the fiber surface can be coated, and that the fiber on the back of the carrier substrate is not coated at all.

2.2. Fabrication Approaches

To determine the most appropriate manufacturing process, we evaluated two different approaches to fabricate TFTs on fibers: The direct fabrication of devices on nylon and glass fibers using standard semiconductor manufacturing equipment [18], and the transfer of TFTs, fabricated on flat and thin substrates, to different fibers, and yarns [24]. During the direct fabrication process the fibers were loaded into the deposition tools by wrapping them around the carrier or using only short ($\approx$6 cm) fibers attached to a carrier.

2.2.1. Direct Fabrication

Direct fabrication was performed on nylon and glass fibers. The schematic process flow is illustrated in Figure 2. Depending on the material, fibers were cleaned using water, acetone, IPA, and sonication. Next, a Cr bottom gate was electron beam evaporated, here the sample was tilted and rotated to ensure a uniform coating of the curved surface. The bottom gate was then insulated by the deposition of a dielectric material. First we used atomic layer deposition (ALD) at 150 °C to grow 100 nm of Al_2O_3. In case of the glass fibers this resulted in an insulating layer, but the high surface roughness of nylon prevented the formation of a pinhole free layer on the nylon fibers. Since ALD is not suitable for the deposition of thicker layers, the nylon fibers were insulated by depositing a 1 μm thick film of parylene. Subsequent to the insulation of the gate, 30 nm of amorphous IGZO was deposited using a radio frequency (RF) magnetron sputtering process based on a ceramic $InGaZnO_4$ target and a pure Ar sputtering atmosphere at a pressure of 2 mTorr. The fabrication process was finalized by the deposition of the source and drain contacts. 10 nm of titanium, acting a adhesion layer, and 75 nm of gold were electron beam evaporated. Structuring of all layers was done using a shadow mask. This is because of the geometry of the fibers, and also due to the limited chemical resistance of nylon fibers. Here, low resolution shadow masks were hand cut from aluminum foil, whereas high resolution ($\approx$100 μm) shadow masks were etched from a polyimide foil structured using conventional lithography [25].

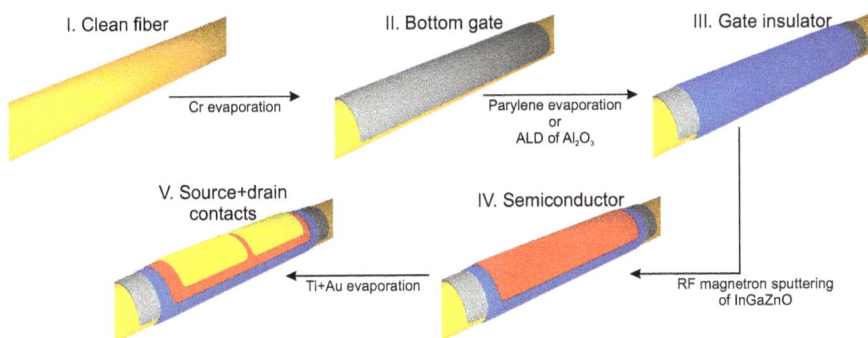

Figure 2. Direct fabrication process flow: Deposition techniques and materials used to manufacture oxide semiconductor thin-film transistors (TFTs) directly on cylindrical fibers. Layer structuring is done by shadow masks.

2.2.2. Transfer Fabrication

Another possibility to overcome the process related limitations caused by the mechanical, chemical and geometrical properties of the different fibers is to fabricate TFTs on a conventional flexible substrate and then transfer them onto a fiber or yarn. This approach was evaluated by fabricating passivated IGZO based bottom gate inverted staggered TFTs on a Si wafer covered with a spin coated 400 nm Polyvinyl alcohol (PVA) sacrificial layer and an evaporated 1 μm thin parylene membrane. The TFTs itself were fabricated by evaporating 35 nm Cr, insulated by an ALD deposited 25 nm Al_2O_3 layer, acting as bottom gate; RF sputtering of 15 nm amorphous IGZO; and the evaporation of 60 nm Au (here, an underlying 15 nm thick Ti layer acts as adhesion layer) as source and drain contacts. Furthermore an additional 25 nm Al_2O_3 layer is used as back-channel passivation. All layers were structured by standard optical lithography. The detailed fabrication process is described elsewhere [24]. After the fabrication, the PVA sacrificial layer is dissolved in water, and the resulting free standing electronic membrane can then be cut and transferred to a fiber. Nylon fibers with radii of 500 μm and 250 μm as well as yarns are used as final substrate. Here the low thickness of the parylene membrane ensures that even the small bending radii caused by wrapping the transistors around a fiber with diameter 250 µm, cannot cause mechanical strain larger than 0.5 %. The reason for this is the direct proportionality between substrate thickness and strain induced by bending. This in return guaranties the full functionality of the transistors. The transfer process is visualized in Figure 3a. To promote the adhesion between the parylene and the nylon, a commercial two component polymercaptan/epoxy adhesive was used. The surface tension of the adhesive also prevented any wrinkling of the parylene membrane. Figure 3b illustrates the structure of the resulting functionalized fibers.

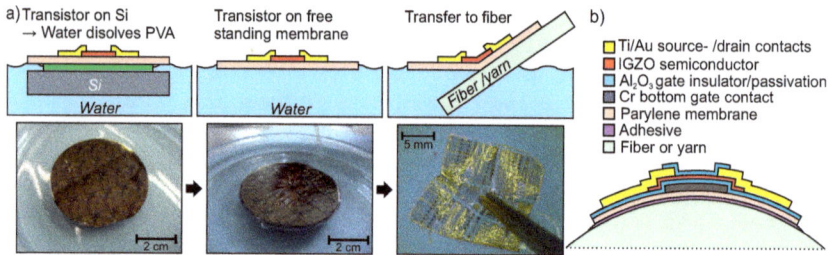

Figure 3. (a) Schematic process flow of the transfer fabrication approach. Here, standard lithography was used to fabricate TFTs on a parylene membrane attached to a standard silicon wafer. Subsequently the TFTs are detached from the wafer by dissolving a corresponding sacrificial layer. Finally the TFTs are transferred to a fiber. (b) Layer structure and materials of the resulting passivated bottom gate inverted staggered InGaZnO (IGZO) TFTs on a fiber or yarn.

3. Results and Discussion

Electrical characterization of the fiber TFTs was performed inside a shielded probe station under ambient conditions using a Keysight B1500A parameter analyzer. Performance parameters were extracted using the Shichman-Hodges equations to model the field effect transistor drain current in the saturation regime [26].

3.1. Directly Fabricated TFTs

The IGZO TFTs, directly fabricated on nylon and glass fibers, are presented in Figure 4. Multiple TFTs have been fabricated on a single fiber, where a common gate was used for all TFTs on one fiber.

3.1.1. TFTs on Polymer Fibers

Figure 4a,b show a photograph and the associated V_{GS}-I_D transfer characteristic of the nylon fiber TFTs. As mentioned above the main obstacle concerning the TFT fabrication on nylon fibers is the high surface roughness of nylon. To effectively insulate the gate from the transistor channel it was necessary to deposit a 1 µm thick parylene layer as gate insulator. This in combination with the low dielectric constant of parylene (3.06) [27] lead to a very low gate capacitance of $\approx 27\,\mu F\,m^{-2}$. Hence, the resulting TFTs exhibit only a low on-off current ratio of $\approx 3 \times 10^2$ even if the gate-source voltage is swept between $-30\,V$ and $47.5\,V$. At the same time it has to be mentioned that even at high voltages like this, the gate current stays below $10^{-9}\,A$. Nevertheless, under the applied gate-source voltages the TFTs are only operated in the subthreshold regime, which excludes the extraction of any meaningful quantitative performance parameters. These results show that the direct fabrication of TFTs on a commercial nylon fiber seems possible. Nevertheless the problems associated to the surface roughness, such as the required thickness of the gate insulator, and hence the high operation voltages, exclude any useful application as long as no better dielectric is found.

3.1.2. TFTs on Glass Fibers

To reduce the operation voltage of the fiber TFTs the gate capacitance has to be increased. Since the deposition of significantly thinner gate insulators is not possible on the employed nylon, glass fibers have been used to prove the concept. The smooth surface and higher temperature resistance of glass allowed the fabrication of functional TFTs using only 100 nm of Al_2O_3, exhibiting a dielectric constant of 9.5, as gate insulator. Figure 4c displays photographs and micrographs of the resulting transistors. The corresponding transfer and output characteristics of a representative TFT are shown in Figure 4d,e, respectively. The transistor operated in depletion mode and exhibits a threshold voltage of -12.5 V, a field effect mobility of $3 \, cm^2 \, V^{-1} \, s^{-1}$, an on-off current ratio of 10^4, and a maximum transconductance of 1.7 µS. Compared to the nylon fiber transistors, these performance parameters show a significant improvement, nonetheless in particular the very negative threshold voltage is not desirable. This is because, for wearable applications, enhancement mode transistors operating at voltages below 5 V are preferred. The reason for the negative threshold voltage is the lack of a back channel passivisation, and the fact that all process steps are performed at room temperature (hence there is no intentional or unintentional annealing of the semiconductor). It is expected that the deposition of an additional Al_2O_3 passivation layer would be beneficial, but structuring and precise alignment of small contact holes on the source and drain contacts using a shadow mask is challenging (the performed structuring of the gate insulator is significantly less demanding). At the same time, fabrication of passivated TFTs using the transfer approach, described in the next paragraph, is easily possible.

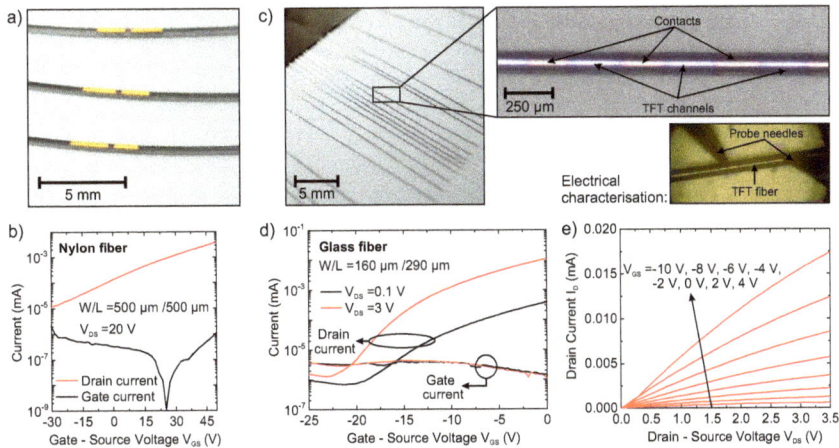

Figure 4. TFTs directly fabricated on fibers: (**a**) Photograph of TFTs on a 500 µm diameter Nylon fiber fabricated using 1 µm parylene as gate insulator. (**b**) Corresponding transistor transfer characteristic. (**c**) Photograph and micrographs of TFTs fabricated on a 125 µm diameter glass fiber fabricated using 100 nm atomic layer deposition (ALD) deposited Al_2O_3 as gate insulator. Corresponding transfer (**d**), and output (**e**) transistor characteristics.

3.2. Transferred TFTs

Figure 5 presents transistors on nylon fibers and yarns fabricated using the transfer approach. A micrograph of a functionalized nylon fiber with a diameter of 250 µm is shown in Figure 5a. A representative TFT has been characterized directly before and after it was transferred to the fiber. The transfer characteristic of the transistor measured on the silicon carrier wafer (Figure 5b) and when permanently attached to the fiber (Figure 5c) shows that the transistor operates in the enhancement mode. On the fiber, the TFTs exhibits a threshold voltage of 0.85 V (-0.1 V compared to the measurement before transfer), a field effect mobility of $7.2\,cm^2\,V^{-1}\,s^{-1}$ (+4 %), an on-off current ratio of 10^7 (×8), and a maximum transconductance of 52.1 µS (+7.2 %). The improvement of the transistor performance is associated to tensile mechanical strain induced by bending the TFT around the fiber. In case of IGZO TFTs this strain increased the effective carrier mobility and decreases the threshold voltage [28]. The corresponding output characteristic of the same transistor measured on the fiber is plotted in Figure 5d, here a clear current saturation effect is visible.

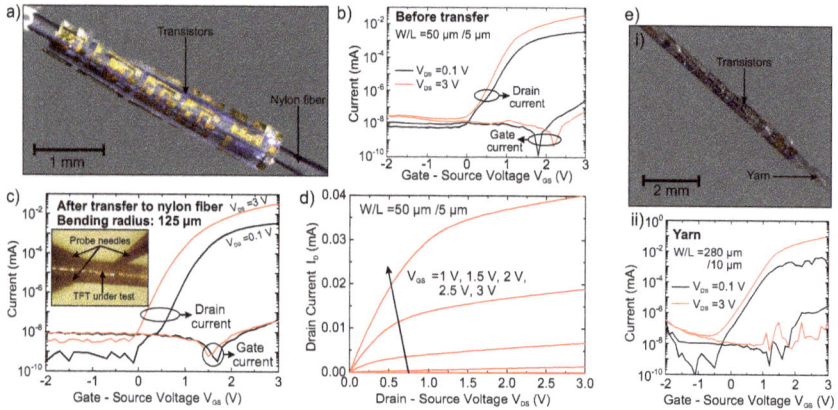

Figure 5. TFTs fabricated on fibers and yarns using the transfer approach: (**a**) Micrograph of TFTs on a 250 µm diameter Nylon fiber. (**b**) The transistor transfer characteristics shown in (**b**) and (**c**) compare the TFT performance measured before and after the TFT was released form the silicon fabrication substrate and transferred to the fiber [the inset in (**c**) illustrates contacting the electronic fiber]. (**d**) Corresponding output characteristic. (**e**) Micrograph [i] and measured transfer characteristic [ii] of a TFT on a yarn.

In addition to the use of nylon fibers, Figure 5e illustrates that it is also possible to fabricate functionalized yarns using the transfer approach. Here, a transistor on a multi-thread yarn with a variable diameter between 100 µm and 250 µm is shown. The corresponding TFT transfer characteristic is used to extract the performance parameters. The noise visible in the measurement of the linear transistor regime (V_{DS} = 0.1 V) is caused by the uneven and soft surface of the yarn. This soft surface prevents the formation of a reliable and stable contact between the device and the probe needles, which also leads to an increase of the contact resistance. On the yarn the transistor exhibits a threshold voltage of 0.8 V, a field effect mobility of $4.6\,cm^2\,V^{-1}\,s^{-1}$, an on-off current ratio of 10^6, and a maximum transconductance of 93 µS, which confirms the full electronic functionality of the TFT.

4. Conclusions

We compared different fibers, yarns and thin-film manufacturing technologies and evaluated their suitability for the fabrication of TFTs on textile fibers with diameters down to 125 µm. The direct fabrication of bottom gate transistors based on amorphous IGZO on nylon and glass fibers is possible. Here, the chemical properties, and the surface roughness of commercial textile fibers degrade the performance of these transistors which makes it difficult to achieve low voltage operation and high flexibility simultaneously. At the same time, fabricating transistors on nylon fibers and yarn using a transfer approach makes it possible to use standard fabrication technologies and to realize functionalized fibers compatible with large scale textile manufacturing equipment. In this respect, transferring IGZO TFTs from a flat substrate to an arbitrary fiber or yarn results in high performance transistors with field effect mobilities up to $7.2\,cm^2\,V^{-1}\,s^{-1}$. Furthermore, an additional advantage of the transfer process is that the length of the fibers is virtually unlimited. Consequently, as demonstrated in Figure 6, the presented technology can be used to realize smart textiles based on active electronic devices which are indistinguishable from the textile fabric itself. Here, the TFTs stay fully functional. Nevertheless, it has to be mentioned that the shown fiber was manually woven into the textile (weft direction) because the available weaving machine feeds fibers though small loops, this potentially destroys the TFTs on the fiber surface (this issue could be addressed by employing an additional structured encapsulation, e.g. made from parylene). At the same time, we already demonstrated that conducive yarns integrated in the warp direction and connected to flexible IGZO TFTs (using conductive epoxy) can be used to form a bus structure inside the textile and to contact the woven transistors [29]. This will contribute to the development and large scale production of future support systems, unobtrusively integrated into industrial fabrics or clothing for sports, safety, and healthcare applications.

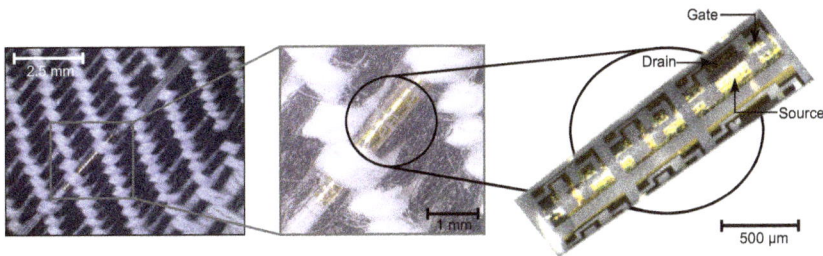

Figure 6. Textile integrated thin-film transistors: IGZO TFTs on Nylon fiber with a diameter of 500 μm are integrated into a commercial textile. The electronic fiber replaces a weft direction cotton yarn.

Acknowledgments: We would like to thank our former student Raoul Guggenheim for his great contribution. This research was support by the Swiss national science foundation (SNF) under the nano-terra grant 3D-SensTex: grant No.: 530803.

Author Contributions: Niko Münzenrieder and Christian Vogt developed the idea and designed the experiments, Niko Münzenrieder and Christian Vogt performed the experiments, Niko Münzenrieder, Christian Vogt, Giuseppe Cantarella, Luisa Petti and Lars Büthe analyzed the data, Niko Münzenrieder wrote the paper with input from all authors, Gerhard Tröster supervised the research.

Conflicts of Interest: The authors declare no conflict of interest.

References

1. Nathan, A.; Ahnood, A.; Cole, M.T.; Lee, S.; Suzuki, Y.; Hiralal, P.; Bonaccorso, F.; Hasan, T.; Garcia-Gancedo, L.; Dyadyusha, A.; et al. Flexible electronics: The next ubiquitous platform. *Proc IEEE* **2012**, *100*, 1486–1517.
2. Amft, O.; Tröster, G. On-body sensing solutions for automatic dietary monitoring. *IEEE Pervasive Comput.* **2009**, *8*, 62–70.
3. Cherenack, K.; van Pieterson, L. Smart textiles: Challenges and opportunities. *J. Appl. Phys.* **2012**, *112*, 091301.
4. Zysset, C.; Kinkeldei, T.; Cherenack, K.; Tröster, G. Woven electronic textiles: An enabling technology for health-care monitoring in clothing. In Proceedings of the UbiComp'10, Copenhagen, Denmark, 26–29 September 2010; pp. 1–4.
5. Codau, T.C.; Onofrei, E.; Bedek, G.; Dupont, D.; Cochrane, C. Embedded textile heat flow sensor characterization and application. *Sens. Actuators A Phys.* **2015**, *235*, 131–139.
6. Giovanelli, D.; Farella, E. Force Sensing Resistor and Evaluation of Technology for Wearable Body Pressure Sensing. *J. Sens.* **2016**, *2016*, doi:10.1155/2016/9391850.
7. Castano, L.M.; Flatau, A.B. Smart fabric sensors and e-textile technologies: A review. *Smart Mater. Struct.* **2014**, *23*, 053001.
8. Pu, X.; Li, L.; Song, H.; Du, C.; Zhao, Z.; Jiang, C.; Cao, G.; Hu, W.; Wang, Z.L. A Self-Charging Power Unit by Integration of a Textile Triboelectric Nanogenerator and a Flexible Lithium-Ion Battery for Wearable Electronics. *Adv. Mater.* **2015**, *27*, 2472–2478.

9.   Dai, M.; Xiao, X.; Chen, X.; Lin, H.; Wu, W.; Chen, S. A low-power and miniaturized electrocardiograph data collection system with smart textile electrodes for monitoring of cardiac function. *Australas. Phys. Eng. Sci. Med.* **2016**, *39*, 1029–1040.

10. Locher, I.; Kirstein, T.; Tröster, G. Temperature profile estimation with smart textiles. In Proceedings of the International Conference on Intelligent textiles, Smart clothing, Well-being, and Design, Tampere, Finland, 19–20 September 2005; p. 8.

11. Lee, J.B.; Subramanian, V. Organic transistors on fiber: A first step towards electronic textiles. In Proceedings of the IEEE International Electron Devices Meeting 2003 (IEDM'03 Technical Digest), Washington, DC, USA, 8–10 December 2003; pp. 199–202.

12. Hamedi, M.; Forchheimer, R.; Inganäs, O. Towards woven logic from organic electronic fibres. *Nat. Mater.* **2007**, *6*, 357–362.

13. Zeng, W.; Shu, L.; Li, Q.; Chen, S.; Wang, F.; Tao, X.M. Fiber-based wearable electronics: A review of materials, fabrication, devices, and applications. *Adv. Mater.* **2014**, *26*, 5310–5336.

14. Cherenack, K.; Zysset, C.; Kinkeldei, T.; Münzenrieder, N.; Tröster, G. Wearable Electronics: Woven Electronic Fibers with Sensing and Display Functions for Smart Textiles. *Adv. Mater.* **2010**, *22*, 5071.

15. Zysset, C.; Kinkeldei, T.; Münzenrieder, N.; Petti, L.; Salvatore, G.; Tröster, G. Combining electronics on flexible plastic strips with textiles. *Text. Res. J.* **2013**, *83*, 1130–1142.

16. Kinkeldei, T.; Denier, C.; Zysset, C.; Münzenrieder, N.; Tröster, G. 2D Thin Film Temperature Sensors Fabricated onto 3D Nylon Yarn Surface for Smart Textile Applications. *Res. J. Text. Appar.* **2013**, *17*, 16–20.

17. Nomura, K.; Ohta, H.; Takagi, A.; Kamiya, T.; Hirano, M.; Hosono, H. Room-temperature fabrication of transparent flexible thin-film transistors using amorphous oxide semiconductors. *Nature* **2004**, *432*, 488–492.

18. Munzenrieder, N.; Petti, L.; Zysset, C.; Salvatore, G.; Kinkeldei, T.; Perumal, C.; Carta, C.; Ellinger, F.; Troster, G. Flexible a-IGZO TFT amplifier fabricated on a free standing polyimide foil operating at 1.2 MHz while bent to a radius of 5 mm. In Proceedings of the 2012 IEEE International the Electron Devices Meeting (IEDM), San Francisco, CA, USA, 10–13 December 2012; pp. 96–99.

19. Petti, L.; Münzenrieder, N.; Vogt, C.; Faber, H.; Büthe, L.; Cantarella, G.; Bottacchi, F.; Anthopoulos, T.D.; Tröster, G. Metal oxide semiconductor thin-film transistors for flexible electronics. *Appl. Phys. Rev.* **2016**, *3*, 021303.

20. Issum, B.V.; Chamberlain, N. The free diameter and specific volume of textile yarns. *J. Text. Inst. Trans.* **1959**, *50*, T599–T623.

21. Zysset, C.; Nasseri, N.; Büthe, L.; Münzenrieder, N.; Kinkeldei, T.; Petti, L.; Kleiser, S.; Salvatore, G.A.; Wolf, M.; Tröster, G. Textile integrated sensors and actuators for near-infrared spectroscopy. *Opt. Express* **2013**, *21*, 3213–3224.

22. Service, R.F. Patterning electronics on the cheap. *Science* **1997**, *278*, 383–384.

23. Hegemann, D.; Amberg, M.; Ritter, A.; Heuberger, M. Recent developments in Ag metallised textiles using plasma sputtering. *Mater. Technol.* **2009**, *24*, 41–45.

24. Salvatore, G.A.; Münzenrieder, N.; Kinkeldei, T.; Petti, L.; Zysset, C.; Strebel, I.; Büthe, L.; Tröster, G. Wafer-scale design of lightweight and transparent electronics that wraps around hairs. *Nat. Commun.* **2014**, *5*, 1–8.

25. Kinkeldei, T.; Munzenrieder, N.; Zysset, C.; Cherenack, K.; Tröster, G. Encapsulation for flexible electronic devices. *IEEE Electron Device Lett.* **2011**, *32*, 1743–1745.

26. Shichman, H.; Hodges, D.A. Modeling and simulation of insulated-gate field-effect transistor switching circuits. *IEEE J. Solid-State Circuits* **1968**, *3*, 285–289.

27. Kondo, M.; Uemura, T.; Matsumoto, T.; Araki, T.; Yoshimoto, S.; Sekitani, T. Ultraflexible and ultrathin polymeric gate insulator for 2 V organic transistor circuits. *Appl. Phys. Express* **2016**, *9*, 061602.

28. Münzenrieder, N.; Cherenack, K.; Tröster, G. The Effects of Mechanical Bending and Illumination on the Performance of Flexible IGZO TFTs. *Trans. Electron Devices IEEE* **2011**, *58*, 2041–2048.

29. Zysset, C.; Munzenrieder, N.; Kinkeldei, T.; Cherenack, K.; Troster, G. Woven active-matrix display. *IEEE Trans. Electron Devices* **2012**, *59*, 721–728.

technologies

MDPI

Article

Dynamic Gesture Recognition Using a Smart Glove in Hand-Assisted Laparoscopic Surgery

Lidia Santos [1,*], Nicola Carbonaro [2,3], Alessandro Tognetti [2,3], José Luis González [1], Eusebio de la Fuente [1], Juan Carlos Fraile [1] and Javier Pérez-Turiel [1]

[1] Instituto de las Tecnologías Avanzadas de la Producción (ITAP), University of Valladolid, 47011 Valladolid, Spain; jossan@eii.uva.es (J.L.G.); efuente@eii.uva.es (E.d.l.F.); jcfraile@eii.uva.es (J.C.F.); turiel@eii.uva.es (J.P.-T.)
[2] Research Centre E. Piaggio, University of Pisa, 56122 Pisa, Italy; nicola.carbonaro@centropiaggio.unipi.it (N.C.); a.tognetti@centropiaggio.unipi.it (A.T.)
[3] Department of Information Engineering, University of Pisa, 56122 Pisa, Italy
* Correspondence: lidia.santos@uva.es; Tel.: +34-983-423-355

Received: 14 November 2017; Accepted: 10 January 2018; Published: 13 January 2018

Abstract: This paper presents a methodology for movement recognition in hand-assisted laparoscopic surgery using a textile-based sensing glove. The aim is to recognize the commands given by the surgeon's hand inside the patient's abdominal cavity in order to guide a collaborative robot. The glove, which incorporates piezoresistive sensors, continuously captures the degree of flexion of the surgeon's fingers. These data are analyzed throughout the surgical operation using an algorithm that detects and recognizes some defined movements as commands for the collaborative robot. However, hand movement recognition is not an easy task, because of the high variability in the motion patterns of different people and situations. The data detected by the sensing glove are analyzed using the following methodology. First, the patterns of the different selected movements are defined. Then, the parameters of the movements for each person are extracted. The parameters concerning bending speed and execution time of the movements are modeled in a prephase, in which all of the necessary information is extracted for subsequent detection during the execution of the motion. The results obtained with 10 different volunteers show a high degree of precision and recall.

Keywords: Hand-Assisted Laparoscopic Surgery (HALS); sensing glove; wearable; collaborative surgical robot; gesture recognition

1. Introduction

One of the most important innovations in surgery over the past three decades has been the advent of minimally invasive surgery (MIS). This technique has revolutionized surgical practice due to its ability to avoid the trauma of traditional open surgery and diminish the possibility of incision-related complications. These benefits also have economic consequences, because they result in a reduction of hospital stay times. However, MIS is technically challenging, because it must be conducted in a very restricted space using micro instruments and endoscopes that dramatically limit the surgeon's perception. In order to gain tactile and force feedback, new technologies and techniques have been introduced over the last few years. One of these novel techniques is hand-assisted laparoscopic surgery (HALS). In HALS, the surgeon inserts a hand in the patient's abdomen through a small incision via a pressurized sleeve while operating a surgical tool with the other hand. Although this approach is slightly more invasive for the patient, it is still a MIS intervention, and has been proved especially advantageous in some types of operations, such as colon and colorectal cancer surgery [1].

However, HALS has a major shortcoming. As the surgeon is holding the tissue with the inserted hand and a micro instrument with the other, he/she needs the close cooperation of an assistant to manage the endoscope and additional surgical tools when performing surgical maneuvers such as stitching and knot tying. In this paper, we tackle the automation of the tasks performed by the human assistant using, instead, a collaborative robot (Figure 1). This robotic system requires, among other important issues, a simple communication scheme capable of recognizing the surgeon's direct orders given by the hand inserted in the abdominal cavity.

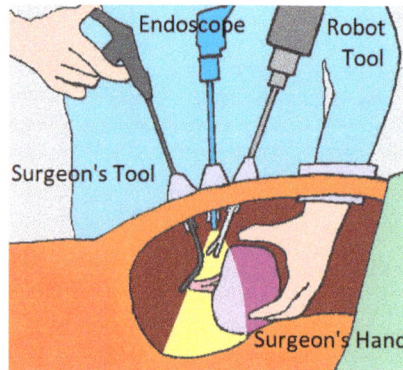

Figure 1. Hand-assisted laparoscopic surgery (HALS) scenario using a robotic assistant.

Previous works on surgical robots can be found in the literature. The first robot systems for laparoscopic surgery were developed to provide more stability and precision to the movements of the surgical tools and endoscopes. They were teleoperated systems that integrated a simple robotic arm with a laparoscopic instrument attached to it [2]. Since then, a number of semi-autonomous robots have been developed and studied to assist the surgeon in the different phases of the operation [3–5].

Autonomous systems require recognition of the surgical gestures made by the surgeon. The use of cameras is an early developed technology to sense hand gestures [6–10], including gloved hand recognition [11], but image processing is always problematic when the scene is under variable illumination or with a cluttered background [12]. In HALS, the variable lighting provided by an endoscope under continuous movement, as well as the difficulty of extracting a permanently blood-stained hand from the internal scenes, prevent the use of this technology. These circumstances are aggravated as the hand would only be partially visible in the images due to the limited viewing field available inside the abdominal cavity.

In order to communicate with the collaborative robot in a natural way, the use of a sensor glove is proposed. A dynamic gesture recognition algorithm has been developed to identify the commands the surgeon gives to the robot with his/her hand inserted in the abdominal cavity. The chosen textile-based motion sensing glove is comfortable, and permits the perfect mobility of the surgeon's hand in the reduced space inside the patient's abdomen. Although the glove used in this study was tailored for a different application (i.e., daily-life monitoring of the grasping activity of stroke patients, as described in [13]) and has a low number of sensors, i.e., three sensors covering the thumb, index, and middle fingers, this wearable device allows the surgeon's hand movements to be monitored. The movement of the fingers is followed by the glove sensors, without limiting the operability. However, some disturbances may appear due to cross-talk between sensors. When the operator tries to move one finger, this can generate a noise signal in another. These disturbances are filtered to avoid misclassification of the surgeon's gesture.

To check the algorithm, 10 tests were performed by 10 different subjects to detect some movements designated in a previous selection phase. Each test consisted of three predefined movements and two additional gestures, which were included in order to demonstrate that the algorithm does not erroneously confuse a movement that is not predefined with a predefined one.

The aim of these tests is to determine whether the developed gesture recognition algorithm can be used to send commands using a sensor glove to a collaborator robot during a HALS with a high degree of precision.

This paper is organized as follows. Section 2 introduces the materials and methodologies used in the experiments that are shown in Section 3. The results are presented in Section 3 and discussed in Section 4. Finally, Section 5 presents the conclusions.

2. Materials and Methods

2.1. Sensing Glove

The sensing glove adopted in this work is made of cotton–lycra, and has three textile goniometers directly attached to the fabric. Figure 2a shows the position of the goniometers on the glove, while Figure 2b show the final prototype of the glove where the goniometers are insulated with an additional layer of black fabric.

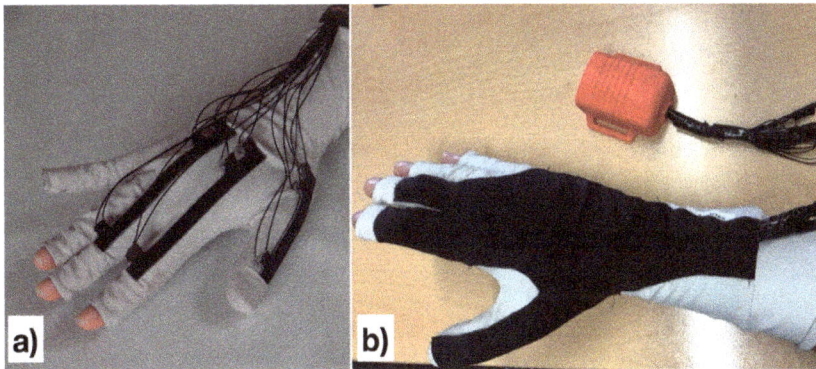

Figure 2. (**a**) The goniometers attached to the glove fabric, (**b**) the sensing glove prototype and the wireless acquisition unit.

The textile goniometers are double layer angular sensors, as previously described in [14,15]. The sensing layers are *knitted piezoresistive fabrics* (KPF) that are made of 75% electro-conductive yarn and 25% Lycra [16,17]. The two KPF layers are coupled through an electrically-insulating stratum (Figure 3a). The sensor output is the electrical resistance difference (ΔR) of the two sensing layers. We demonstrated earlier that the sensor output is proportional to the flexion angle (θ) [14], which is the angle delimited by the tangent planes to the sensor extremities (Figure 3b).

The glove was developed in previous studies to monitor stroke patients' everyday activity to evaluate the outcome of their rehabilitation treatment [13,18]. In [19], the reliable performance of the glove goniometers was demonstrated, and showed errors below five degrees as compared with an optical motion capture instrument during natural hand opening/closing movements. The glove has two KPF goniometers on the dorsal side of the hand to detect the flexion-extension movement of the metacarpal-phalangeal joints of the index and middle fingers. The third goniometer covers the trapezium-metacarpal and the metacarpal-phalangeal joints of the thumb to detect thumb opposition. We conceived this minimal sensor configuration as a tradeoff between grasping recognition and the wearability of the prototype.

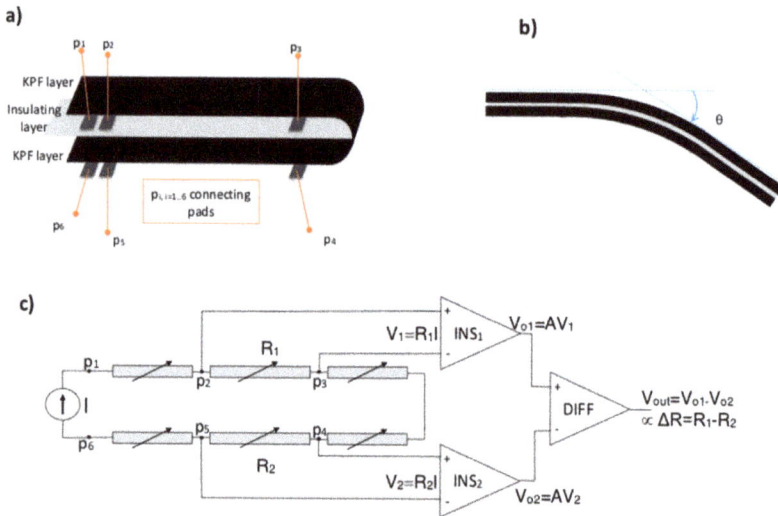

Figure 3. (a) Schematic structure of the knitted piezoresistive fabrics (KPF) goniometer. The black stripes represent the two identical piezoresistive layers, while the gray stripe is the insulating layer; **(b)** The output (ΔR) is proportional to the bending angle (θ) **(c)** KPF goniometer electrical model and block diagram of the electronics front-end. Two instrumentation amplifiers (INS_1 and INS_2) and a differential amplifier (DIFF) produce the output ΔV, which is proportional to ΔR and thus to $\Delta \theta$.

An ad hoc three-channel analog front-end was designed for the acquisition of ΔR from each of the three goniometers (Figure 3c). For each goniometer, the voltages $V_1 = V_{p2} - V_{p3}$ and $V_2 = V_{p5} - V_{p4}$ are measured when a constant and known current I is supplied through p_1 p_6. A high-input impedance stage, consisting of two instrumentation amplifiers (INS_1 and INS_2), measures the voltages across the KPF sensors. These voltages are proportional, through the known current I, to the resistances of the top and bottom layers (R_1 and R_2). A differential amplifier (DIFF) amplifies the difference between the measured voltages, obtaining the final output ΔV, which is proportional to ΔR and θ. Each channel was analogically low-pass filtered (anti-aliasing, cut-off frequency of 10 Hz). The resulting data were digitally converted (sample time of 100 Sa/s) and wirelessly transmitted to a remote PC for storage and further elaboration.

2.2. Algorithm for Movement Detection

The glove will communicate with a collaborative robot to assist during a HALS. The actions to test the collaborative robot take into account the various robotic actions covered by the literature [20], among which are the guidance of the laparoscopic camera for the safe movement of the endoscope [21] or a needle insertion [22], the prediction of the end point [23,24], the knotting and unknotting on suture procedures [25], or grasping and lifting on tissue retraction [26]. Ultimately, we selected three actions to be performed by the collaborative robot: center the image from the endoscope, indicate a place to suture, and stretch the thread to suture. These actions are performed in a cholecystectomy, which is the surgical removal of the gallbladder.

Each of the three robot actions mentioned above is associated with a hand movement to be performed by the surgeon. Therefore, the system must be prepared to unambiguously recognize the different movements defined as commands for the robot in order to prevent it from performing undesirable operations. They will be differentiated by the detection algorithm, which is tested with a protocol.

The protocol includes these three movements, which must be detected as robot commands, and are shown in Table 1 and numbered from 1 to 3. Actions 4 and 5 are introduced to test the developed algorithm. These were selected for their similarity to the movements selected in both the sensor value and motion patterns. As a result, differentiation in advance between the different movements is difficult.

Table 1. Selected movements to be detected.

N°	Initial Posture	Final Posture	Description	Command
1			From initial posture to final posture twice	To center the image from the endoscope.
2			From initial posture to final posture twice.	To indicate a place to suture.
3		-	Initial posture for a defined time.	To indicate to stretch the thread.
4			From initial posture to final posture twice.	-
5			From initial posture to final posture twice.	-

To detect these movements, the developed algorithm analyzes the following parameters: flexion pattern, velocity, execution times, and value provided by the sensor of each finger. To evaluate these parameters, there is a previous phase in which the variables of each movement in each person are examined. This previous stage is required for each person, because the speed and timing of the finger movement is highly variable, as shown in Figure 4.

Once these variables are defined, as explained in later paragraphs, the detection algorithm can identify each of the three movements.

The motion of the index and middle fingers is sensed by the glove. The acquired data is continuously processed by the developed algorithm in order to detect some of the predefined dynamics patterns. Due to the unique textile substrate to which of all the sensors are attached, cross-talk between sensors may appear. This could be observed as a disturbing signal from a finger when the operator tries to move another finger, as shown in Figure 5. These movements are filtered in order to avoid a misclassification.

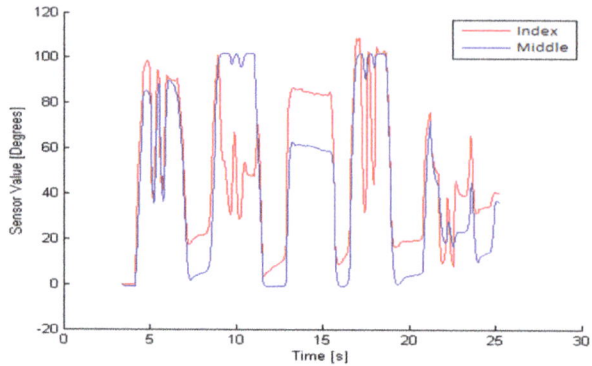

(a)

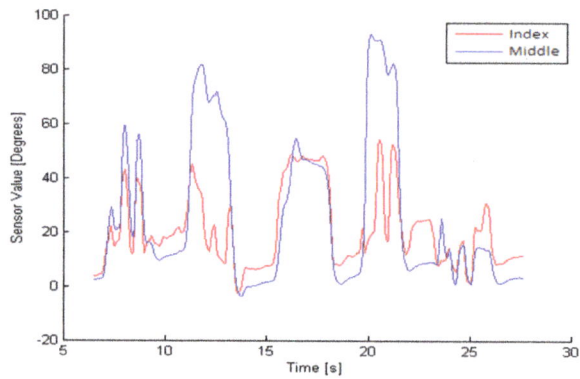

(b)

Figure 4. Sensor values during the performance of the movements represented in Table 1 in the same order. (**a**) person 1, and (**b**) person 2.

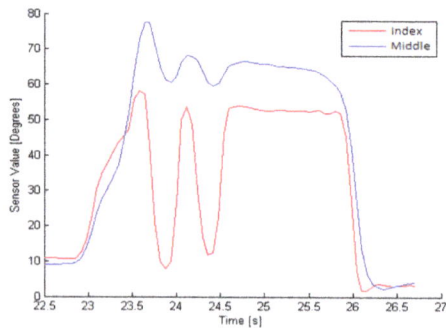

Figure 5. Sensor values during movement 2. There should be no motion in the middle finger, because only the index finger should participate.

Due to the nature of the sensors used, it is possible to determine the degree of flexion being applied to the sensor on the glove. However, movements 4 and 2 could be confused due to their similarity, as shown in Figure 6.

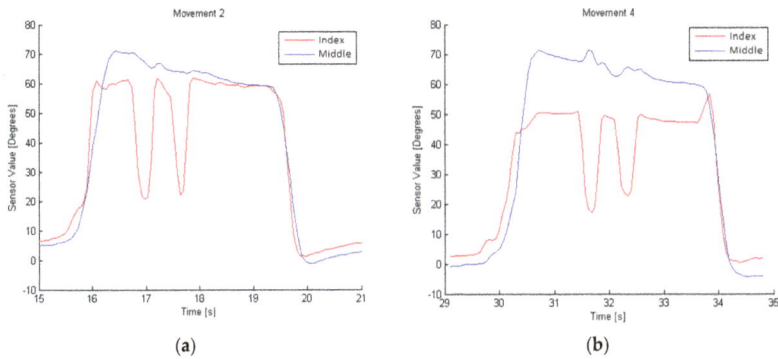

Figure 6. (**a**) Sensor values during movement 2 and (**b**) movement 4, performed by the same person.

Movement 1 can be identified by analyzing the data from the index and middle fingers. Each rise and fall in the glove sensor values corresponds to the flexion and extension movements of the fingers. This movement consists of a descent (called D1) and ascent (A1), followed by another descent (D2) and ascent (A2), as shown in Figure 7. This is the flexion pattern considered for movement 1. The D time and A time are, respectively, the times taken during a descent or ascent.

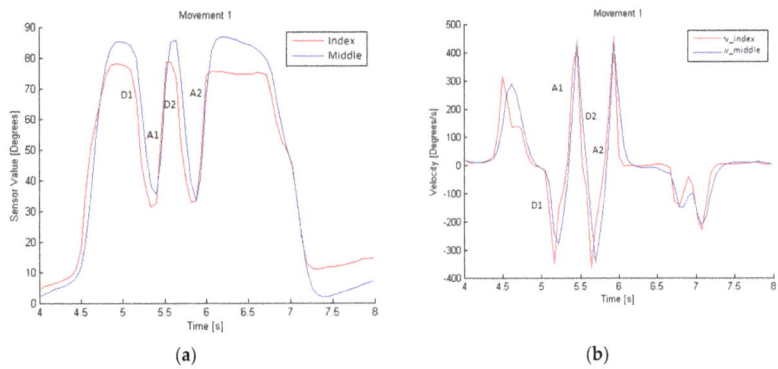

Figure 7. (**a**) Glove sensor data, which are proportional to the flexion of the finger in movement 1; (**b**) Velocity of flexion involved in movement 1.

The flexion velocity involved in this dynamic gesture is higher than the cross-talk ones, as shown in Figure 7b. To establish the typical velocity for this movement, the average and the standard deviation of the velocity along D1 and D2, and A1 and A2, are calculated. This typical velocity, V_{1u}, is the minimum value obtained from the subtraction of the standard deviation from the average in three tests performed by the same person. The minimum time during descents, t_{1Du}, (D1 and D2) and ascents, t_{1Au}, (A1 and A2) is also calculated, and will represent the characteristic ascent and descent execution times of movement 1.

To determine the execution time, t_{1u}, the maximum time in which the whole movement is performed is considered; that is D1, A1, D2, and A2.

The last parameters to be defined are the maximum, x_{max}, and minimum, x_{min}, values of the sensor, which set the thresholds to consider if the obtained values are part of movement 1. They are obtained by analyzing three movement samples from the same person.

With these parameters, shown in Table 2, movement 1 can be defined and differentiated from others, considering the flexion velocity V_e as the instantaneous velocity scanned during the entire movement performed, and the execution time t_e as the time in which the velocity exceeds the velocity threshold.

Table 2. Characterization of defined movements.

Mov.	Finger	Flexion Pattern	Flexion Velocity	Execution Time	D Time	A Time	Sensor Value
1	Index Middle	D1 A1 D2 A2	$\|V_e\| > V_{1u}$	$t_e < t_{1u}$	$t_D > t_{1Du}$	$t_A > t_{1Au}$	$x_{min} < x < x_{max}$
2	Index	D1 A1 D2 A2	$\|V_e\| > V_{2u}$	$t_e < t_{2u}$	$t_D > t_{1Du}$	$t_A > t_{1Au}$	$x_{min} < x < x_{max}$
3	Index Middle	-	-	$t_e > t_{3u}$	-	-	$x_{min} < x < x_{max}$

Using the graphs obtained during the performance of movement 2, as shown in Figure 8, we can conclude that it is necessary to determine the movements of the index and middle finger in order to obtain a definition. The flexion pattern for this movement is D1, A1, D2, and A2 for the index finger, and no movement for the middle finger. The velocity, time of execution, minimum time during descents (D1 and D2) and ascents (A1 and A2), and the sensor value are defined as described in movement 1.

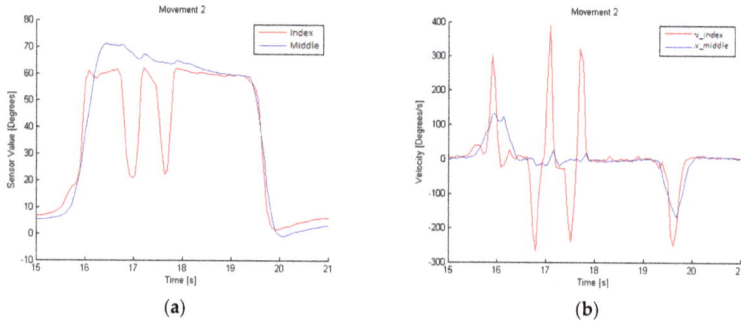

Figure 8. (a) Glove sensor data, which are proportional to the flexion of the finger in movement 2; (b) Velocity of the flexion involved in movement 2.

Movement 3, in Figure 9, differs from the other two in that the velocity must be 0, so it is a static position maintained for a certain time. To identify it, we examine the values of the index and middle finger sensors, which will be proportional to the flexion carried out by the finger with the sensor.

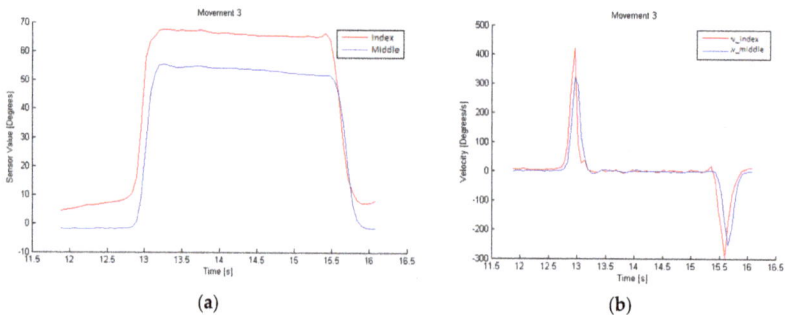

Figure 9. (a) Glove sensor data, which are proportional to the flexion of the finger in movement 3; (b) Velocity of the flexion involved in movement 3.

The algorithm for the detection of defined movements evaluates all of the abovementioned parameters, and detects when one of these movements is executed.

3. Experiments and Results

The test consists of carrying out the movements shown in Table 1 in the same order, as well as performing a flat position between them, in a scenario of experiments (Figure 10). Therefore, the correct order of execution is: flat position, movement 1, flat position, movement 2, flat position, movement 3, flat position, movement 4, flat position, movement 5, flat position.

Figure 10. Scenario of experiments with a pelvitrainer, which simulates the patient's abdomen. The collaborative robot is holding the endoscope. The sensing glove is partially viewed on the screen.

Movements 1, 2, and 3 have been selected to be detected by the algorithm, while movements 4 and 5 were introduced to prove that they are not detected in the same manner as the three selected ones. The two newly introduced movements are similar to movements 1 and 2, but there are small differences between them.

First, the data are collected from the glove. Then, they are analyzed by the algorithm to detect the movements that will be interpreted as commands for the robot. These orders are then sent to the collaborative robot.

The test has been carried out by 10 people, 10 times. Ten right-handed volunteers (five men, five women) completed the test. All of the participating people in these research activities gave informed consent for the experiments. No one reported physical limitations that would affect their skill in performing the task.

The characteristic parameters of each movement are calculated from three tests performed by the same person. These parameters are characteristic of each person, so 10 sets of patterns have been obtained for each type of movement, one per person.

Movements 1, 2, and 3 must be detected by the algorithm, while movements 4 and 5 should not be classified as selected movements. As shown in Table 3, movement 1 was detected with a precision—the percentage of positive predictions that were correct—of 0.99, and a recall—the percent of the positive cases recognized—of 0.98. Movement 4 was identified as movement 1 only 1% of the time. On the other hand, movement 2 was detected with a precision of 0.73 and a recall of 0.87. Movement 4 was recognized as movement 2 33% of the time. Movement 5 was never mistaken with other movements, and movement 3 was detected with a precision of 1.0 and a recall of 0.97.

Table 3. Total results.

Total		Actual Movement					Precision	Recall	F_1-Score
		1	**2**	**3**	**4**	**5**			
	1	98	0	0	1	0	0.99	0.98	0.98
Predicted Movement	2	0	87	0	33	0	0.73	0.87	0.79
	3	0	0	97	0	0	1.00	0.97	0.98

F_1 as scored for movements 1 and 3 is 0.98, while for movement 2 it was considerably lower, 0.79.

4. Discussion

Movement 4 is detected as movement 2 or 1 because of their similarity, as explained in the previous sections. Despite the study of different patterns, times, and speeds, movement 4 is detected as movement 2 35% of the time. Whenever movement 3 has not been detected, this was due to an insufficient time in the static position.

Reviewing the results, it can be concluded that the effectiveness of the algorithm depends largely on the person performing the test as shown in Appendix A. Results with surgeons are expected to be better, because they have greater motor skills, considering their specific training [27]. Tests have shown that the newly developed algorithm can adequately identify the three movements defined in a series of different continuous movements. Movement recognition is precise, because identification is based not only on the initial and final pose, but also on intermediate positions and speeds that are continuously analyzed to determine whether their pattern is analogous to the model. Different filters are also introduced to make the dynamic gesture recognition algorithm more reliable. The patterns obtained with the sensing glove present sufficient information as to be robustly identified, and prevent failures in those cases where the positions are similar to those of the model, but the execution speed of the movement is different.

One of the purposes of this study was to test the validity of our non-specific glove to demonstrate the possibility of using this kind of device, and define the specification for a HALS-dedicated textile glove for use in future studies. In future works, glove-based hand motion sensing could be fused with other sensing modalities, such as artificial vison, to make the system more robust.

5. Conclusions

Most current surgical robots are not suitable for HALS operations. Its teleoperated nature prevents its application in these operations where the surgeon is in direct contact with the patient. In this scenario, it is necessary to have a robot co-worker that cooperates closely with the surgeon in order to emulate the interaction with a human assistant. A natural communication interface between surgeon and robot is crucial in this context. This paper tackles the design of a dynamic gesture recognition algorithm using a sensor glove that identifies the commands given by the surgeon's hand inside the patient's abdominal cavity. Three different dynamic gestures have been predefined to: point the robot where to suture, order it to focus the endoscope, and stretch the thread. All of these tasks present automatic procedures in the literature to carry them out. The algorithm designed to recognize these gestures analyzes continuously the timing and the bending speed of the index and middle fingers, and it tries to match them with some of the patterns previously recorded by a particular operator.

The experiments conducted with 10 different volunteers show a good recognition rate and time performance. However, considering its application in surgical operations, there is room for improvement. Although this study has considered the option of the sensing glove, another hand motion sensor would need to be added in order to make the system completely reliable. Furthermore, other important issues such as safety or electromagnetic compatibility should be addressed in future works.

Acknowledgments: This research has been partially funded by the Spanish State Secretariat for Research, Development and Innovation, through project DPI2013-47196-C3-3-R.

Author Contributions: Alessandro Tognetti and Nicola Carbonaro conceived and designed the sensing glove. Lidia Santos carried out the experiments. Lidia Santos processed and analyzed the data. Eusebio de la Fuente, José Luis González, Alessandro Tognetti and Nicola Carbonaro gave advice and discussion. Lidia Santos, Eusebio de la Fuente, Alessandro Tognetti and Nicola Carbonaro wrote the paper. Juan Carlos Fraile, Javier Turiel, Alessandro Tognetti and Nicola Carbonaro supervised the entire work. All authors have read and approved the final manuscript.

Conflicts of Interest: The authors declare no conflict of interest.

Appendix A

Results for each volunteer.

Table A1. Results Volunteer 1.

Volunteer 1		Actual Movement					Precision	Recall	F_1-Score
		1	2	3	4	5			
Predicted Movement	1	10	0	0	0	0	1.00	1.00	1.00
	2	0	8	0	0	0	1.00	0.80	0.89
	3	0	0	10	0	0	1.00	1.00	1.00

Table A2. Results Volunteer 2.

Volunteer 2		Actual Movement					Precision	Recall	F_1-Score
		1	2	3	4	5			
Predicted Movement	1	10	0	0	0	0	1.00	1.00	1.00
	2	0	10	0	2	0	0.83	1.00	0.91
	3	0	0	10	0	0	1.00	1.00	1.00

Table A3. Results Volunteer 3.

Volunteer 3		Actual Movement					Precision	Recall	F_1-Score
		1	2	3	4	5			
Predicted Movement	1	10	0	0	0	0	1.00	1.00	1.00
	2	0	8	0	2	0	0.80	0.80	0.80
	3	0	0	10	0	0	1.00	1.00	1.00

Table A4. Results Volunteer 4.

Volunteer 4		Actual Movement					Precision	Recall	F_1-Score
		1	2	3	4	5			
Predicted Movement	1	10	0	0	0	0	1.00	1.00	1.00
	2	0	9	0	5	0	0.64	0.90	0.75
	3	0	0	9	0	0	1.00	0.90	0.95

Table A5. Results Volunteer 5.

Volunteer 5		Actual Movement					Precision	Recall	F_1-Score
		1	2	3	4	5			
Predicted Movement	1	10	0	0	0	0	1.00	1.00	1.00
	2	0	9	0	2	0	0.82	0.90	0.86
	3	0	0	10	0	0	1.00	1.00	1.00

Table A6. Results Volunteer 6.

Volunteer 6		Actual Movement					Precision	Recall	F_1-Score
		1	2	3	4	5			
Predicted Movement	1	10	0	0	0	0	1.00	1.00	1.00
	2	0	8	0	7	0	0.53	0.80	0.64
	3	0	0	10	0	0	1.00	1.00	1.00

Table A7. Results Volunteer 7.

Volunteer 7		Actual Movement					Precision	Recall	F_1-Score
		1	2	3	4	5			
Predicted Movement	1	10	0	0	0	0	1.00	1.00	1.00
	2	0	8	0	9	0	0.47	0.80	0.59
	3	0	0	10	0	0	1.00	1.00	1.00

Table A8. Results Volunteer 8.

Volunteer 8		Actual Movement					Precision	Recall	F_1-Score
		1	2	3	4	5			
Predicted Movement	1	10	0	0	1	0	0.91	1.00	0.95
	2	0	10	0	0	0	1.00	1.00	1.00
	3	0	0	9	0	0	1.00	0.90	0.95

Table A9. Results Volunteer 9.

Volunteer 9		Actual Movement					Precision	Recall	F_1-Score
		1	2	3	4	5			
Predicted Movement	1	8	0	0	0	0	1.00	0.80	0.89
	2	0	7	0	0	0	1.00	0.70	0.82
	3	0	0	9	0	0	1.00	0.90	0.95

Table A10. Results Volunteer 10.

Volunteer 10		Actual Movement					Precision	Recall	F_1-Score
		1	2	3	4	5			
Predicted Movement	1	10	0	0	0	0	1.00	1.00	1.00
	2	0	10	0	6	0	0.63	1.00	0.77
	3	0	0	10	0	0	1.00	1.00	1.00

References

1. Jayne, D.G.; Thorpe, H.C.; Copeland, J.; Quirke, P.; Brown, J.M.; Guillou, P.J. Five-year follow-up of the Medical Research Council CLASICC trial of laparoscopically assisted versus open surgery for colorectal cancer. *Br. J. Surg.* **2010**, *97*, 1638–1645. [CrossRef] [PubMed]
2. LaRose, D.; Taylor, R.H.; Funda, J.; Eldridge, B.; Gomory, S.; Talamini, M.; Kavoussi, L.; Anderson, J.; Gruben, K. A Telerobotic Assistant for Laparoscopic Surgery. *IEEE Eng. Med. Biol. Mag.* **1995**, *14*, 279–288.
3. Bauzano, E.; Garcia-Morales, I.; del Saz-Orozco, P.; Fraile, J.C.; Muñoz, V.F. A minimally invasive surgery robotic assistant for HALS-SILS techniques. *Comput. Methods Programs Biomed.* **2013**, *112*, 272–283. [CrossRef] [PubMed]
4. Kim, K.Y.; Song, H.S.; Suh, J.W.; Lee, J.J. A novel surgical manipulator with workspace-conversion ability for telesurgery. *IEEE/ASME Trans. Mechatron.* **2013**, *18*, 200–211. [CrossRef]

5. Estebanez, B.; del Saz-Orozco, P.; García-Morales, I.; Muñoz, V.F. Multimodal Interface for a Surgical Robotic Assistant: Surgical Maneuvers Recognition Approach. *Rev. Iberoam. Autom. Inform. Ind.* **2011**, *8*, 24–34. [CrossRef]
6. Song, Y.; Demirdjian, D.; Davis, R. Continuous body and hand gesture recognition for natural human-computer interaction. *Int. Jt. Conf. Artif. Intell.* **2015**, *2*, 4212–4216. [CrossRef]
7. Ganokratanaa, T.; Pumrin, S. The vision-based hand gesture recognition using blob analysis. In Proceedings of the 2017 International Conference on Digital Arts, Media and Technology (ICDAMT), Chiang Mai, Thailand, 1–4 March 2017; pp. 336–341.
8. Asadi-Aghbolaghi, M.; Clapes, A.; Bellantonio, M.; Escalante, H.J.; Ponce-Lopez, V.; Baro, X.; Guyon, I.; Kasaei, S.; Escalera, S. A Survey on Deep Learning Based Approaches for Action and Gesture Recognition in Image Sequences. In Proceedings of the 2017 12th IEEE International Conference on Automatic Face & Gesture Recognition (FG 2017), Washington, DC, USA, 30 May–3 June 2017; pp. 476–483.
9. Alon, J.; Athitsos, V.; Yuan, Q.; Sclaroff, S. Simultaneous localization and recognition of dynamic hand gestures. In Proceedings of the 2005 WACV/MOTIONS '05 Volume 1. Seventh IEEE Workshops on Application of Computer Vision, Breckenridge, CO, USA, 5–7 January 2005; pp. 254–260.
10. Suryanarayan, P.; Subramanian, A.; Mandalapu, D. Dynamic hand pose recognition using depth data. In Proceedings of the 2010 20th International Conference on Pattern Recognition, Istanbul, Turkey, 23–26 August 2010; pp. 3105–3108.
11. Nguyen, B.P.; Tay, W.L.; Chui, C.K. Robust Biometric Recognition from Palm Depth Images for Gloved Hands. *IEEE Trans. Hum. Mach. Syst.* **2015**, *45*, 799–804. [CrossRef]
12. Lu, Z.; Chen, X.; Li, Q.; Zhang, X.; Zhou, P. A hand gesture recognition framework and wearable gesture-based interaction prototype for mobile devices. *IEEE Trans. Hum. Mach. Syst.* **2014**, *44*, 293–299. [CrossRef]
13. Lorussi, F.; Carbonaro, N.; De Rossi, D.; Paradiso, R.; Veltink, P.; Tognetti, A. Wearable Textile Platform for Assessing Stroke Patient Treatment in Daily Life Conditions. *Front. Bioeng. Biotechnol.* **2016**, *4*, 28. [CrossRef] [PubMed]
14. Tognetti, A.; Lorussi, F.; Mura, G.; Carbonaro, N.; Pacelli, M.; Paradiso, R.; Rossi, D. New generation of wearable goniometers for motion capture systems. *J. Neuroeng. Rehabil.* **2014**, *11*, 56. [CrossRef] [PubMed]
15. Tognetti, A.; Lorussi, F.; Carbonaro, N.; de Rossi, D. Wearable goniometer and accelerometer sensory fusion for knee joint angle measurement in daily life. *Sensors* **2015**, *15*, 28435–28455. [CrossRef] [PubMed]
16. Pacelli, M.; Caldani, L.; Paradiso, R. Performances evaluation of piezoresistive fabric sensors as function of yarn structure. In Proceedings of the 2013 35th Annual International Conference of the IEEE Engineering in Medicine and Biology Society (EMBC), Osaka, Japan, 3–7 July 2013; pp. 6502–6505.
17. Pacelli, M.; Caldani, L.; Paradiso, R. Textile piezoresistive sensors for biomechanical variables monitoring. In Proceedings of the 2006 International Conference of the IEEE Engineering in Medicine and Biology Society, New York, NY, USA, 30 August–3 September 2006; pp. 5358–5361.
18. Tognetti, A.; Lorussi, F.; Carbonaro, N.; De Rossi, D.; De Toma, G.; Mancuso, C.; Paradiso, R.; Luinge, H.; Reenalda, J.; Droog, E.; et al. Daily-life monitoring of stroke survivors motor performance: The INTERACTION sensing system. In Proceedings of the 2014 36th Annual International Conference of the IEEE Engineering in Medicine and Biology Society, Chicago, IL, USA, 26–30 August 2014; pp. 4099–4102.
19. Carbonaro, N.; Mura, G.D.; Lorussi, F.; Paradiso, R.; De Rossi, D.; Tognetti, A. Exploiting wearable goniometer technology for motion sensing gloves. *IEEE J. Biomed. Health Inform.* **2014**, *18*, 1788–1795. [CrossRef] [PubMed]
20. Kranzfelder, M.; Staub, C.; Fiolka, A.; Schneider, A.; Gillen, S.; Wilhelm, D.; Friess, H.; Knoll, A.; Feussner, H. Toward increased autonomy in the surgical OR: Needs, requests, and expectations. *Surg. Endosc.* **2013**, *27*, 1681–1688. [CrossRef] [PubMed]
21. Moustris, G.P.; Hiridis, S.C.; Deliparaschos, K.M.; Konstantinidis, K.M. Robust feature tracking in the beating heart for a robotic-guided endoscope. *Int. J. Med. Robot.* **2011**, *7*, 375–392. [CrossRef] [PubMed]
22. Wen, R.; Tay, W.L.; Nguyen, B.P.; Chng, C.B.; Chui, C.K. Hand gesture guided robot-assisted surgery based on a direct augmented reality interface. *Comput. Methods Programs Biomed.* **2014**, *116*, 68–80. [CrossRef] [PubMed]

23. Weede, O.; Mönnich, H.; Müller, B.; Wörn, H. An intelligent and autonomous endoscopic guidance system for minimally invasive surgery. In Proceedings of the 2011 IEEE International Conference on Robotics and Automation, Shanghai, China, 9–13 May 2011; pp. 5762–5768.
24. Staub, C.; Osa, T.; Knoll, A.; Bauernschmitt, R. Automation of tissue piercing using circular needles and vision guidance for computer aided laparoscopic surgery. In Proceedings of the 2010 IEEE International Conference on Robotics and Automation, Anchorage, AK, USA, 3–7 May 2010; pp. 4585–4590.
25. Shi, H.F.; Payandeh, S. Real-time knotting and unkotting. In Proceedings of the 2007 IEEE International Conference on Robotics and Automation, Roma, Italy, 10–14 April 2007; pp. 2570–2575.
26. Patil, S.; Alterovitz, R. Toward automated tissue retraction in robot-assisted surgery. In Proceedings of the 2010 IEEE International Conference on Robotics and Automation, Anchorage, AK, USA, 3–7 May 2010; pp. 2088–2094.
27. Richard, K.; Reznick, H.M. Changes in the wind. *N. Engl. J. Med.* **2006**, *355*, 2664–2669.

technologies

MDPI

Article

Perspective Preserving Solution for Quasi-Orthoscopic Video See-Through HMDs

Fabrizio Cutolo *, Umberto Fontana and Vincenzo Ferrari

Information Engineering Department, University of Pisa, 56122 Pisa, Italy; umbertofontana93@gmail.com (U.F.); vincenzo.ferrari@unipi.it (V.F.)
* Correspondence: fabrizio.cutolo@endocas.unipi.it; Tel.: +39-050-995-689

Received: 18 December 2017; Accepted: 10 January 2018; Published: 13 January 2018

Abstract: In non-orthoscopic video see-through (VST) head-mounted displays (HMDs), depth perception through stereopsis is adversely affected by sources of spatial perception errors. Solutions for parallax-free and orthoscopic VST HMDs were considered to ensure proper space perception but at expenses of an increased bulkiness and weight. In this work, we present a hybrid video-optical see-through HMD the geometry of which explicitly violates the rigorous conditions of orthostereoscopy. For properly recovering natural stereo fusion of the scene within the personal space in a region around a predefined distance from the observer, we partially resolve the eye-camera parallax by warping the camera images through a perspective preserving homography that accounts for the geometry of the VST HMD and refers to such distance. For validating our solution; we conducted objective and subjective tests. The goal of the tests was to assess the efficacy of our solution in recovering natural depth perception in the space around said reference distance. The results obtained showed that the quasi-orthoscopic setting of the HMD; together with the perspective preserving image warping; allow the recovering of a correct perception of the relative depths. The perceived distortion of space around the reference plane proved to be not as severe as predicted by the mathematical models.

Keywords: video see-through head-mounted displays; orthoscopy; perspective-preserving homography; stereo fusion

1. Introduction

Augmented reality (AR) systems based on head-mounted displays (HMDs) intrinsically provide the user with an egocentric viewpoint and represent the most ergonomic and efficient solution for aiding manual tasks performed under direct vision [1]. AR HMDs are commonly classified according to the AR paradigm they implement: video see-through (VST) or optical see-through (OST). In binocular VST HMDs, the view of the real world is captured by a pair of stereo cameras rigidly anchored to the visor with an anthropometric interaxial distance. The stereo views of the world are presented onto the HMD after being coherently combined with the virtual content [2].

By contrast, in OST HMDs, the user's direct view of the world is preserved. The fundamental OST paradigm in HMDs is still the same as that described by Benton (e.g., Google Glass, Microsoft HoloLens, Epson Moverio, Lumus Optical) [3]. The user's own view of the real world is herein augmented by projecting the virtual information on a beam combiner and then into the user's line of sight [4].

Although the OST HMDs were once at the leading edge of the AR research, their degree of adoption and diffusion slowed down over the years due to technological and human-factor limitations. Just to mention a few of them: the presence of a small augmentable field of view, the reduced brightness offered by standard LCOS micro displays, the perceptual conflicts between the 3D real world and the 2D virtual image and the need for accurate and robust eye-to-display calibrations [5].

Some of the technological limitations, like the small field of view, are being and will be likely overcome along with the technological progress. The remaining limitations are harder to cope with.

The pixel-wise video mixing technology that underpins the VST paradigm can offer high geometric coherence between virtual and real content. The main reasons for it are: unlike OST displays, the absence of a user-specific eye-to-display calibration routine; the possibility of rendering synchronously real scene and the virtual content, whereas in OST displays there is an intrinsic lag between the immediate perception of the real scene and the appearance of the virtual elements. From a perceptual standpoint, in VST systems the visual experience of both the real and virtual content can be unambiguously controllable by computer graphics, with everything on focus at the same apparent distance from the user. Finally, VST systems are much more suited than OST systems, to rendering occlusions between real and virtual elements or to implementing complex visualization processing modalities that are able to perceptually compensate for the loss of the direct real-world view.

Despite all these advantages, the visual perception of the real world with VST HMDs is adversely affected by various geometric aberrations [6–8]. These geometric aberrations are due to the intrinsic features of cameras and displays (e.g., resolutions limitations and optical distortions) and can be boosted by their relative positioning.

One of the major geometric aberrations typical of VST HMDs is related to the misalignment of viewpoints (parallax) between the capturing cameras and the user's perspective through the display (i.e., non-orthoscopic setup). The parallax between capturing camera and user's viewpoint produces distortion into the patterns of horizontal and vertical binocular disparities and this translates into a distorted perception of space.

To recover proper space perception, researchers have put forward various solutions for implementing claimed parallax-free and orthoscopic VST HMDs [9]. In 1998 Fuchs et al. [10] were the first to propose a parallax-free VST HMD. In that system, a pair of mirrors was used to bring the camera centres in the same location of the nodal point of the wearer's eyes.

In a work published in 2000 [11], a systematic analysis of all the possible distortions in depth perception due to non-rigorous orthostereoscopic configurations was presented. Starting from this comprehensive analysis, the authors pursued the same objective of developing a parallax-free VST HMD by means of a set of mirrors and optical prisms whose goal was to align the optical axes of the displays to those of the two cameras. However, also this solution was characterized by a divergence from the conditions of orthostereoscopy in terms of an offset of approximately 30 mm between camera centre and exit pupil of the display, whose effect the authors claimed to be negligible in terms of perceptive distortions of space.

In 2005 State et al. [12], presented an innovative VST HMD specifically designed to generate zero eye-camera offset. Their system, specifically intended for use in medical applications, was designed and optimized through a software simulator the outputs of which then guided the development of a proof-of-concept prototype, built via rapid prototyping and by assembling off-the-shelf components. In their simulated scenario, the authors properly addressed all the aspects for implementing an orthoscopic VST visor; yet their actual embodiment did not satisfy all those requirements due to the constructive complexities (e.g., it did not comprise any eye tracker). Therefore, their system could provide a parallax-free perception of the reality only for user-specific and constant settings in terms of eye position, inter-pupillary distance and eye convergence.

Finally, in 2009 Bottecchia et al. [13], proposed an orthoscopic monocular prototype of VST HMD in which a computer-based correction of the parallax was mentioned. Unfortunately, the authors then did not provide further details on the way the parallax was resolved via software.

Unfortunately, all the presented solutions were bulky and mostly designed for applications in which the pair of stereoscopic cameras is mounted parallel to each other.

By contrast, for those AR applications in which the user is asked to interact with the augmented scene within personal space (i.e., at distances below 2 m), the stereo camera pair ought to be pre-set at a fixed convergence for ensuring sufficient stereo overlaps and granting proper stereo fusion, i.e., toed-in setup [14–17]. This angle of convergence should be established based on assumptions made on the average working distance. In such configuration, for preserving a natural visual perception of the space (i.e., the conditions of orthostereoscopy) and reduce stereoscopic distortions as keystone distortion and depth-plane curvature, theoretically also the two displays should be physically converged of the same angle [11,18]. Yet, this requirement cannot be fulfilled from a practical standpoint and this has implications on the ability of the stereoscopic system to recovering natural depth perception.

When VST HMDs with parallel stereo cameras are intended for use in close-range tasks, a valid alternative is represented by the purely software mechanism proposed in [19]. In their solution, the idea was to maximize via software the stereo overlaps by handling dynamically the convergence or the shearing of the display frustum based on a heuristic estimation of the working distance.

In line with this approach, we here present a method for properly recovering natural stereo fusion of the scene at a predefined distance from the observer in a binocular VST system designed for tasks performed within arm's reach. Our method explicitly takes into consideration the geometry of the setup and the intrinsic parameters of camera and display for computing the appropriate plane-induced homography between the image planes of the stereo cameras and those of their associated displays. On each side, such perspective preserving homography is used for consistently warping the image grabbed by the camera before rendering it onto the corresponding display. This solution, yields a parallax-free perception of the reference plane and, together with the quasi-orthoscopic setup of the VST HMD, manages to recover almost entirely the natural perception of depth in the space around the reference distance. The selection of the reference plane for the homography for a specific use case is based on assumptions made on the average working distance.

For validating our approach, we took advantage of the hybrid nature of a custom-made see-through HMD [20], which supports both video and optical see-through modalities, for drawing an experimental setup whose goal was twofold. First goal was to assess, under OST view, the resulting monoscopic displacement between real features and synthetic ones at various depths around the one taken as reference for the estimation of the homography. The second goal of the tests was to evaluate quantitatively whether and how such displacements affected the perception of the relative depths in the scene under VST view. To this end, we eventually performed preliminary subjective tests aimed at measuring the accuracy in perceiving relative depths through the VST HMD.

2. Materials and Methods

This section is structured as follows. Section 2.1 provides a detailed description of the binocular hybrid video-optical see-through HMD used in this study. Section 2.2 outlines the geometry of the homography induced by a plane that yields a consistent perspective-preserving image warping of the camera frames. Section 2.3 briefly contains a short description of the AR software framework running on the HMD. Finally, Section 2.4 introduces the methodology adopted for validating the method.

2.1. Binocular Hybrid Video/Optical See-Through HMD

In a previous study, we presented a novel approach for the development of stereoscopic AR HMDs able to offer the benefits of both the video and the optical see-through paradigms [20]. The hybrid mechanism was made possible by means of a pair of electrically-driven LC shutters (FOS model by LC-Tec [21]) mounted ahead of the waveguides of a OST HMD, opportunely modified for housing a pair of stereo cameras. The transition between the unaided (OST) and the camera-mediated (VST) view of the real scene is allowed by acting on the transmittance of the electro-optical shutter. As in the first prototype, the hybrid VST/OST HMD is based on a reworked version of a commercial binocular OST HMD (DK-33 by LUMUS [22]). The optical engine of the visor features a 1280×720 resolution, a horizontal FoV (hFoV) of $35.2°$ and a vertical FoV (vFoV) of $20.2°$. The stereo camera pair is composed by two Sony FCB-MA13 cameras equipped with a $1/2.45''$ CMOS sensor; the cameras are extremely compact in size ($16.5 \times 10.3 \times 18.0$ mm) and have the following characteristics: horizontal FoV = $53°$, vertical FoV = $29°$ and frame rate of 30 fps at 1920×1080 resolution.

The key differences between the previous prototype and the one that we used in this study are: stereo camera placement and orientation (Figure 1).

As for the stereo camera placement, to pursue a quasi-orthostereoscopic view of the scene under VST modality here we opted for a setup featuring an anthropometric interaxial distance (~65 mm), hence we mounted the pair of cameras on the top of the two waveguides. To the same end and as previously done in [1,19,23], we opted for a parallel stereo camera setting. Indeed, in AR visors specifically designed for close-up tasks as ours, a toed-in stereo camera setting would undoubtedly widen the area of possible stereo overlaps [17,24]. Yet this configuration, if not coupled with a simultaneous convergence of the optical display axes, would also distort the horizontal and vertical patterns of binocular disparities between the stereo frames. This fact would go against the achievement of a quasi-orthostereoscopic VST HMD and it is deemed to lead to significant distortions in absolute and relative depth perception [25,26].

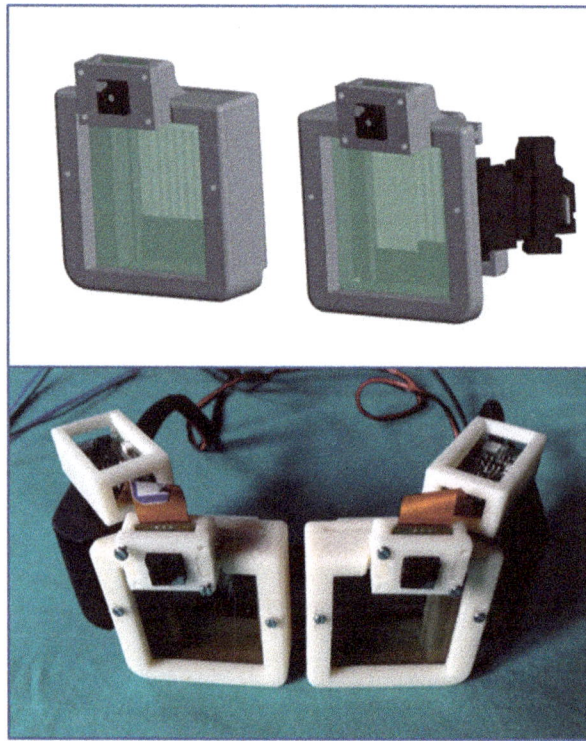

Figure 1. On the top: CAD schematic of the hybrid video/optical see-through head-mounted display comprising the supports for the electro-optical shutters and a pair of stereo cameras mounted on top of the two waveguides of an OST HMD. On the bottom: the HMD.

2.2. Perspective Preserving Planar Homography

This section describes the procedure followed for computing the perspective preserving homography. The goal was to find the geometric relation between two perspective visions of a planar scene placed at a pre-defined distance. With reference to Figure 2 and to the equations below, from now on we shall consider the following convention of variables and symbols:

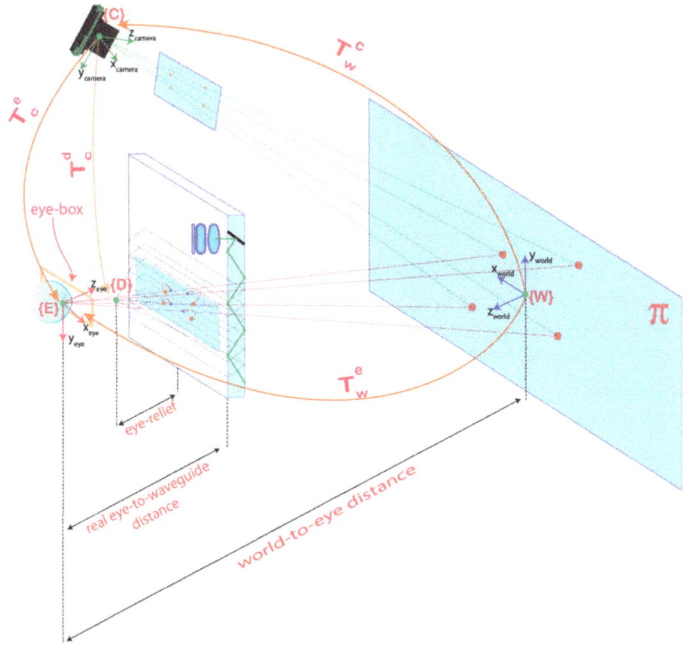

Figure 2. Geometry of the perspective preserving homography induced by a reference plane placed at a pre-defined distance from the eye.

- The homography transformations H_W^D and H_W^C, which relate respectively the points of the reference plane π in the world to their projections onto the image planes of both the display and the camera:

$$\begin{aligned} \lambda_d x_D &= H_W^D \, X_W \\ \lambda_c x_C &= H_W^C \, X_W \end{aligned} \quad \forall \, X_W \in \pi \tag{1}$$

where the points are expressed in homogeneous coordinates and where λ_c and λ_D are generic scale factors due to the equivalence of homogeneous coordinates rule.
- The distance $d^{D \to \pi}$ between the vertex of the display frustum (D) and the reference plane π.
- The eye relief, which represents the fixed distance between D and the eyepiece lens of the display.
- The eye-box (or eye motion box), which consists of that range of allowed eye's positions, at a pre-established eye-relief distance, from where the full image produced by the eyepiece of the display is visible.
- R_C^D and $\vec{t}_C^{\,D}$, which are respectively the rotation matrix and the translation vector between camera reference system (CRS) and display reference system (DRS).
- K_C and K_D, which are the intrinsic matrixes of camera and display. K_C encapsulates the camera intrinsic parameters and it is computed by using the Zhang's method [27] implemented within the camera calibrator tool of MATLAB. K_D encapsulates the parameters of the near-eye display's frustum and it is approximately derived from the specifics of the HMD as follows [28]. We derived the focal length of the display (f) by using the factory specifics of the horizontal and vertical FoV of the display. In our ideal pinhole camera model of the display, the focal length was set equal on both x-axis and y-axis ($f_x = f_y$), meaning the pixels of the display were considered as perfectly

square. As coordinates of the principal point (C_u and C_v) we considered exactly half of the display resolution (C_u = Width/2 = 640, C_v = Height/2 = 360). In summary, we assumed:

$$K_D = \begin{bmatrix} f_x & 0 & C_u \\ 0 & f_y & C_v \\ 0 & 0 & 1 \end{bmatrix} = \begin{bmatrix} W/\left(2 \cdot \tan \frac{hF_0V}{2}\right) & 0 & W/2 \\ 0 & H/\left(2 \cdot \tan \frac{vF_0V}{2}\right) & H/2 \\ 0 & 0 & 1 \end{bmatrix} \tag{2}$$

- H_C^D, which is the perspective preserving homography, induced by a fixed plane π placed at distance $d^{D \to \pi}$ from D.
- $\vec{n}$, which is the normal unit vector to the plane π.

The sought homography transformation H_C^D, describes the point-to-point relation between camera viewpoint and user's viewpoint, such that:

$$\lambda x_D = H_C^D \left(R_C^D, t_C^D, K_C, K_D, \pi\right) x_C \tag{3}$$

The parenthesis means that the homography H_C^D is a function of respectively: the relative pose between camera reference system and display reference system (R_C^D, t_C^D), the intrinsic parameters of camera and display (K_C, K_D) and the position and orientation of the reference plane in the scene (π).

For referring everything to the display we can compute H_D^C and inverting the result afterwards (see Equation (4)).

H_C^D is described by a matrix $H_C^D \in \Re^{3 \times 3}$ and it is function of the pose between camera frustum and display frustum and of the two intrinsic matrixes as follows (for referring everything to the display we have computed H_C^D by inverting H_D^C) [29–31]:

$$H_C^D = H_D^{C-1} = \left(K_C \left(R_D^C + \frac{\vec{t}_D^C \cdot \vec{n}^T}{d^{D \to \pi}} \right) K_D^{-1} \right)^{-1} \tag{4}$$

The homography transformation (4) is only valid on a fixed plane, perpendicular to the optical axis of the display and placed at a predefined distance ($d^{D \to \pi}$). If the plane under observation is different or if the observed scene is not planar, its perceived view (through-the-waveguide view) does not match with the rendered image on the display (i.e., direct view and VST view are not orthoscopically registered).

Another important aspect to consider is the actual position of the nodal point of the user's eye (E) with respect to the DRS. This brings about changes in the variables plugged in Equation (3): in the previous equations, we assumed that the nodal point of the eye (eye centre) was located exactly at the vertex of the display frustum (i.e., $E \equiv D$ or $\vec{t}_D^C \equiv \vec{t}_E^C$). Unfortunately, this is hardly the case in reality. In addition, we must consider the optical properties of the display eyepiece (i.e., eye relief, eye-box, virtual or focal plane position) (Figure 3). These properties play a role in the way in first approximation the non-ideal eye placement in the display reference system ($E \neq D$) affects the elements of K_D.

In summary, Equation (4) becomes:

$$H_C^E = H_E^{C-1} = \left(K_C \left(R_D^C + \frac{\vec{t}_E^C \cdot \vec{n}^T}{d^{E \to \pi}} \right) \widetilde{K_D}^{-1} \right)^{-1} \tag{5}$$

where $\vec{t}_E^C$ ought to be known and where the intrinsic matrix of the display $\widetilde{K_D}$ is different from the original K_D. In view of these considerations, an orthoscopic alignment is attained in theory only if we could determine with absolute accuracy the user's eyes position in the HMD's eyepiece reference system (i.e., DRS). Indeed, in Equation (3) the pose between eye and camera assumes a key role.

Unfortunately, the eye position in the DRS and consequently in the CRS, varies according to how the HMD is worn and is dependent on the user's facial shape (e.g., inter-pupillary distance).

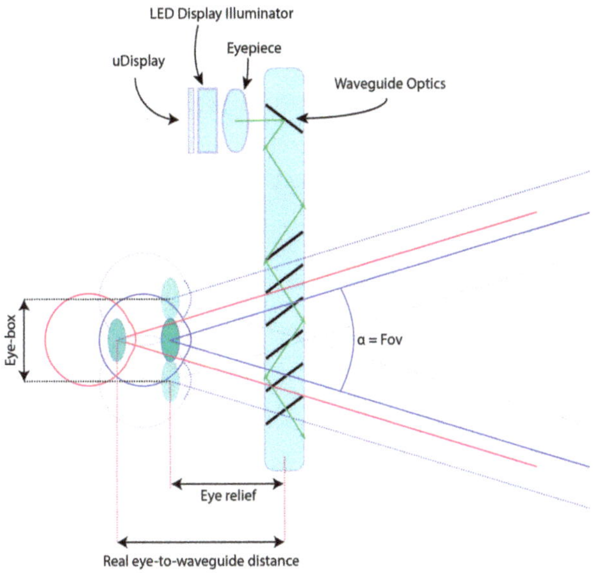

Figure 3. Optical properties of the near-eye display.

Since our HMD did not comprise any eye-tracker to calibrate the stereo cameras to the user's inter-pupillary distance and since we did not perform any specific display calibration (for precisely determining $\widetilde{K_D}$), in our tests we determined an approximation of the homography $\widetilde{H_C^E}$ as follows. We asked the user to wear the HMD and observe under OST modality a target checkerboard placed orthogonally to his own viewpoint at approximately the distance of the homography plane π (Figure 4). We then performed, in real time, multiple refinements of the initial homography H_C^D by means of an additional translational homography $(\widetilde{h})$ whose role was to align the user's views of the checkerboard (real and synthetic). With this additional homography we intended to compensate for the uncertainties in defining the actual position of the eye E with respect to D. In our method, we only considered the effect of translational movements along the x and y-axis (parallel to the image plane). Thus, to a first approximation, we excluded the effect of a non-perfect placement of the user's eye centres at the eye-relief (i.e., we assumed: eye relief = real eye-to-waveguide distance). The relation between the approximated homography $\widetilde{H_C^E}$ and the ideal H_C^D then becomes:

$$\widetilde{H_C^E} = \widetilde{h}\, H_C^D = \begin{bmatrix} 1 & 0 & x_p \\ 0 & 1 & y_p \\ 0 & 0 & 1 \end{bmatrix} H_C^D \tag{6}$$

In conclusion, unlike the method proposed by Tomioka et al. [30], we estimate the user-specific homography uniquely by means of design and calibration data. The homography is then refined to embody the effects of the intrinsic parameters of the displays and of the non-ideal eyes placement in the HMD so as to be adapted to the subject's interpupillary distance (IPD).

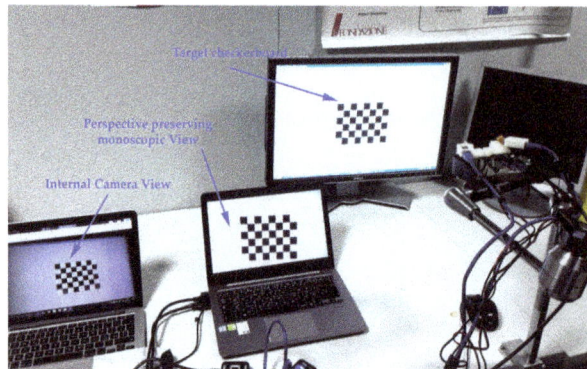

Figure 4. Experimental setup for measuring the monoscopic disparities. The same target checkerboard was used also for the user-specific homography refinement.

2.3. Software Application

For validating our method, we developed a dedicated software application whose main goal was to manage the camera frames as follows. Camera frames are first grabbed and opportunely undistorted for eliminating the non-linearities due to radial distortions. The undistorted frames are warped according to the perspective preserving homography. The warped frames are rendered onto the background of a stereoscopic scenegraph that is finally screened onto the binocular HMD. For this application, we did not add any properly registered virtual content to the scenegraph, as our objective was uniquely to perform perceptual studies on how depth perception was retrieved under VST view.

The application was created in the form of a single executable file with shared libraries all built in C++, following the same logic of the AR software framework previously developed in [32]. As for the library managing the rendering of the scenegraph, we used the open-source library for 3D computer graphics and visualization Visualization Toolkit (VTK), version 7.0.0 [33]. As regards the machine vision routines, needed for processing the camera frames before rendering them onto the background of the scene-graph, we adopted the open-source software library OpenCV 3.1 [34].

2.4. Tests

The proposed solution combines a perspective preserving warping mechanism with a quasi-orthoscopic setting of the VST HMD. The goal of our tests was to assess the efficacy of such solution in recovering the natural perception of depth in the space around a pre-established distance from the observer. We grouped the tests into two basic categories: tests for measuring the patterns of on-image disparities, under OST modality, between real features and HMD-mediated ones and tests for assessing objectively and subjectively the depth estimation accuracy under VST modality.

2.4.1. Test 1: Measure of on-Image Displacements between Direct View and VST View

For measuring the patterns of monoscopic disparities between direct view and VST view, we used the experimental setup showed on Figure 4. The on-image displacements between real features and HMD-mediated ones were measured by means of an additional video camera (Sony FCB-MA13) placed approximately at the ideal eye's position (internal camera). As target scene, we used a standard checkerboard of size 160 × 120 mm (with square size 20 mm) that was displayed on an external monitor. The internal camera was able to capture two views of the target scene: a direct view and a VST view (Figure 5). The corners of the checkerboard could be robustly detected through a Matlab's function for corners detection. The on-image displacements or monoscopic disparities between the image coordinates of the real and VST views of the corner points were so easily determined. The real poses of the target planes with respect to the internal camera were estimated by solving a standard perspective-n-point problem and knowing the 3D-2D point correspondences.

We repeated such measurements at various depths around the one taken as reference for the estimation of the homography (plane π). The range of depths for which the disparities were measured was: (250–650) mm.

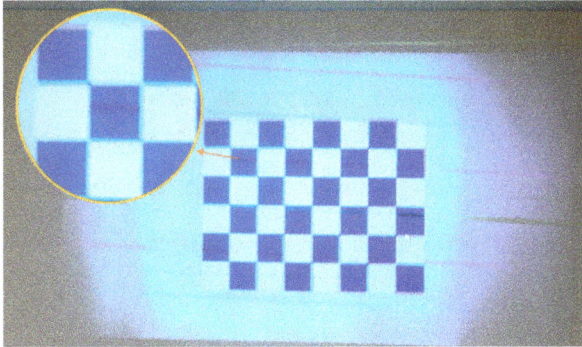

Figure 5. On-image displacements between direct view and VST view of a target scene. The test images were grabbed by an additional video camera placed approximately at the eye relief point of the eyepiece of the HMD.

2.4.2. Test 2: Assessment of Depth Perception through Objective and Subjective Measures

For assessing the degree of accuracy in depth estimation under VST view, we conducted 2 different sets of measurements. At first, we computed the resulting angles of retinal disparities yielded by the monoscopic displacements on both views; these binocular disparities can provide a quantitative estimation of the uncertainty in detecting the relative depths between objects due to the non-ideally orthoscopic setting of the VST HMD. Secondly, we performed a preliminary user study aimed at assessing the accuracy in perceiving relative depths at different distances within personal space (within 1.2 m). Before the session of tests, the homography was refined under OST modality to be adapted to the subject's IPD.

In the tests, one participant wearing the HMD under VST modality was asked to estimate the relative depth relations between three objects of same size and colour (three yellow Lego® bricks of size 9.6 × 32 × 16 mm). We engaged only one participant so as to be consistent in terms of user's stereoscopic acuity. The bricks were laid on five different A3 paper sheets (size 297 × 420 mm), each of which provided with demarcation lines indicating different relative depths. In each triplet of demarcation lines, the relative distances between the bricks were decided randomly, with a defined relative distance between two adjacent bricks of 2 mm.

The paper sheets were placed at five different distances from the observer (Figure 6), covering a range of depths of about 900 mm (i.e., from a minimum absolute depth of 300 mm to a maximum depth of 1200 mm). For each position of the paper sheet, the test was repeated four times.

The perceptual tests were all performed keeping the same homography transformation. For all the target planes, the computed homography was referred to a reference plane perpendicular to the optical axis of the display and placed at a distance of 500 mm. The final goal of the tests was indeed to assess on how this aspect would have had a detrimental effect on perceiving relative depths for all the tested distances of the triplets of bricks.

Each paper included ten possible configurations of relative positions between the three bricks, so we tested a total of 4 (n° of sessions) × 5 (n° of paper sheets at different depths) × 10 (n° of configurations of triplets per paper) = 200 configurations of triplets of bricks.

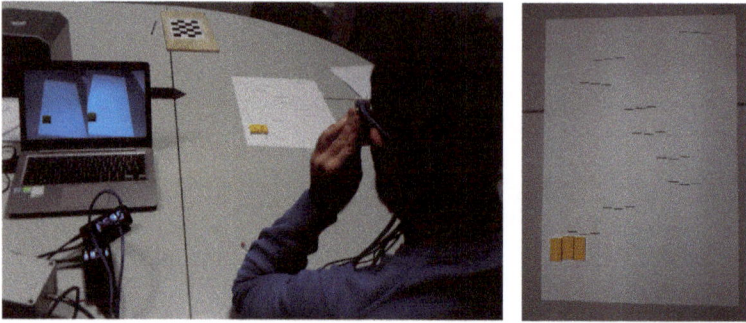

Figure 6. On the left: experimental setup for depth estimation tests. On the right: paper sheet with marked segments. The segments were used for placing the triplets of bricks at different absolute depths from the observer and with random patterns of relative depths among them.

3. Results

This section reports on the results of the two sessions of tests.

3.1. Results of Test 1

We measured the patterns of monoscopic disparities by moving the target checkerboard at ten different positions with respect to the HMD. Each checkerboard contains a set of 35 corner points which results in a total of 350 feature points to be considered in our evaluation. In Figures 7 and 8, the resulting horizontal (h^d) and vertical (v^d) disparities for all the ten positions of the target plane are shown in function of the z-coordinate of the point. The z coordinate of the point is retrieved knowing the pose of the target plane to which they belong. The maps of disparities for each target position are reported in Appendix A.

In relation to the distance from the reference plane π, the vertical disparities follow a steeper increase with respect to the horizontal disparities. This fact directly results from the vertical parallax between CRS and DRS, while the horizontal disparities are only functions of the distance from the reference plane. In Figure 9, we show the horizontal disparities for the points belonging to the six target planes closer to the reference plane. Here the range of depths is: (479–555) mm. In Table 1 the values of the mean and standard deviation of the horizontal disparities are reported for all the target positions.

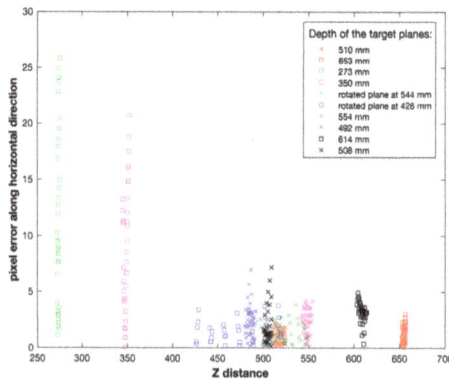

Figure 7. Horizontal monoscopic disparities.

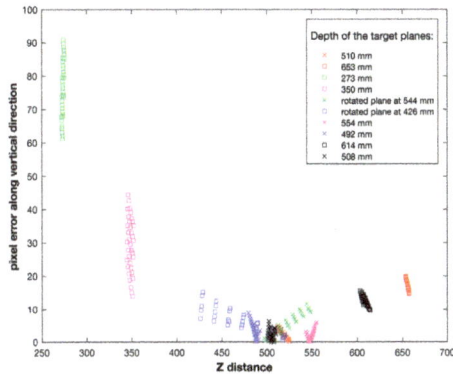

Figure 8. Vertical monoscopic disparities.

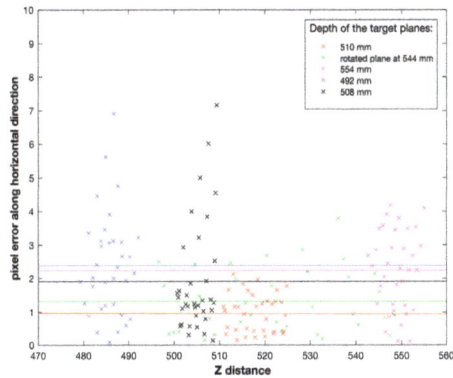

Figure 9. Horizontal monoscopic disparities for the target planes around z = 500 mm. The coloured lines are the mean pixel errors on the u coordinate for each target plane.

Table 1. Mean and standard deviation of horizontal monoscopic disparities between direct-view and VST views of the target planes.

Target Plane Position	Mean Error on u Coordinate (Pixel)	Error's Standard Deviation
273 mm	11.2072	7.6836
350 mm	8.8623	5.6106
426 mm (rotated)	1.5173	1.0251
492 mm	2.3730	1.5282
508 mm	1.9128	1.6844
510 mm	0.9542	0.5848
544 mm (rotated)	1.3085	0.9572
554 mm	2.2297	1.1904
614 mm	3.1658	0.9445
653 mm	1.1834	0.7813

3.2. Results of Test 2

3.2.1. Estimation of Depth Perception under VST View

We here provide a quantitative estimation of the misperception of depth due to the unwanted disparities on both the sides of the HMD. In stereoscopic displays as binocular HMDs, human stereopsis is usually simulated by generating pairs of stereo views with a perceptually consistent amount of horizontal disparities. The relation between the binocular horizontal disparity observed on the 3D display (d_d) and the retinal disparity (d_r) is the following [35]:

$$d_r = 2 \cdot atan \left(\frac{d_d \cdot tan(\alpha/2)}{W} \right) \approx \frac{2 \cdot d_d \cdot tan(\alpha/2)}{W} \tag{7}$$

By plugging $W = 1280$ and $\alpha = 35.2°$ into the equation, we can calculate the minimum angular disparity (or minimum angular resolution) that our HMD is able to provide (for a display disparity $d_d = 1$ pixel): $d_{rmin} \cong 1.7$ arcmin. This value leads to a visual acuity of about half of the average visual acuity in human vision (visual acuity $= d_r{}^{-1}$). The approximated formula of the depth resolution dZ at a distance Z from the observer can be retrieved as follows [18,36,37]:

$$dZ = \frac{Z^2 d_r}{I \cdot K_r} \tag{8}$$

where d_r is expressed in arcmins, I is the observer's IPD and $K_r = 3437.75$ is a constant that relates radians to arcmins. By considering a standard value of $I = 65$ mm, the stereoacuity or depth resolution offered by our HMD at 500 mm is of about 2 mm.

By plugging Equation (8) in Equation (7), we obtain the relation between depth resolution and binocular horizontal disparity:

$$dZ = \frac{2 \cdot Z^2 \cdot d_d \cdot tan(\alpha/2)}{W \cdot I \cdot K_r} \tag{9}$$

where the binocular horizontal disparity can be expressed in terms of image coordinates as follows: $d_d = u_r - u_l$.

If we consider the values of the stereoacuity offered by our HMD at different depths, we are able to compute the ideal density of the homographies from Z_{min} to Z_{max}:

$$Z_i = Z_{i-1} + dZ_i = Z_{i-1} + \frac{Z_i^2 d_r}{I \cdot K_r} \tag{10}$$

For instance, in the range of depths between 250 and 650 mm ($Z_0 = 250$ and $Z_{max} \geq 650$), in theory we would need as much as 350 different homographies in order to stay within the resolution constraints of the HMD. In spite of this and as we explained in Section 2.4.2, the perceptual tests were all performed keeping the same homography transformation for all the target planes, since our goal was to assess on how this aspect would have affected depth perception.

In the first session of tests, we observed how the non-ideally orthoscopic setting of the HMD causes unwanted monoscopic disparities h^d on both the sides of the HMD the further we go from the reference distance. Thus, the horizontal disparity can be written as follows:

$$\tilde{d}_d = u_r \pm \left(h_r^d \right) - \left(u_l \pm h_l^d \right) = d_d \pm 2 \cdot h^d \tag{11}$$

In the equation, we assumed the worst-case scenario, where monocular disparities on both sides add together and they have the same value. In this way, we can estimate the contribution of such disparities to the depth resolution:

$$\widetilde{dZ} = \frac{2 \cdot Z^2 \cdot \left(d_d + 2 \cdot h^d \right) \cdot tan(\alpha/2)}{W \cdot I \cdot K_r} \tag{12}$$

So, the overall depth resolution is affected by the additional disparity contribution brought by the non-ideally orthoscopic setting of the visor. In the range of depths around the plane π ($z = 500$ mm) used for computing the homography, the mean of $\widetilde{dZ}$ was of about 8.8 mm. The value of $\widetilde{dZ}$ is lower if we consider a smaller area at the center of the stereo images, that is where the monoscopic horizontal disparities are not as high.

In Figure 10, the profile of $\widetilde{dZ}$ at various distances from the observer is shown for different values of horizontal disparities. In the figure, we also report the values of $\widetilde{dZ}$ associated to the measured values of h^d for $Z \in [490 - 510]$ mm.

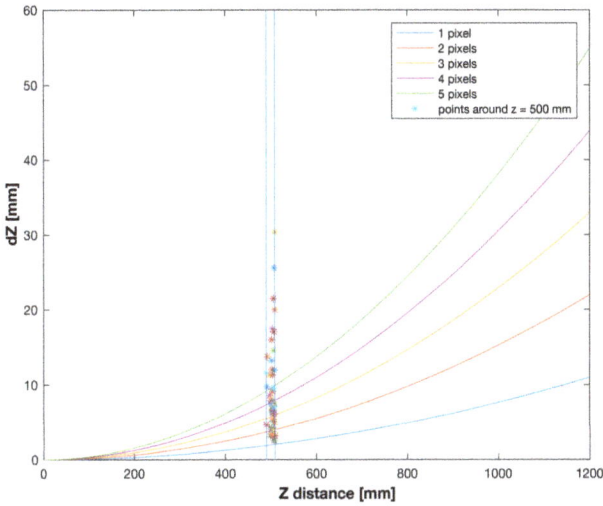

Figure 10. Depth resolution vs distance from the observer with different values of horizontal disparities. The asterisks are the values of $\widetilde{dZ}$ associated to the measured disparities around plane π.

3.2.2. Perceptual Tests

In Table 2, the results of the perceptual tests are reported. We measured the rate of success in terms of proper relative depths estimation among the bricks in all the tested configurations.

The success rate was surprisingly higher than expected (98.5% of total success rate), taking into consideration the fact that the relative depths between the three bricks was between 2 and 4 mm at any distance. In the next session, we shall motivate for the apparent inconsistency between the misperception of relative depths predicted by the mathematical models and the perceived distortion of space experienced by the user during the real use cases.

Table 2. Results of the depth estimation tests.

Distance Range (mm)	First Test Success Rate	Second Test Success Rate	Third Test Success Rate	Fourth Test Success Rate	
~300–~750	100%	100%	100%	100%	
~400–~850	100%	100%	100%	100%	
~500–~950	100%	90%	100%	100%	
~600–~1050	100%	100%	90%	100%	
~700–~1150	100%	100%	90%	100%	
Total Success Rate for each test	100%	98%	96%	100%	Total Success Rate 98.5%

4. Discussion & Conclusions

In this work, we have presented a VST HMD whose geometry violates the rigorous conditions of orthostereoscopy. For properly recovering natural stereo fusion of the scene in a region around a predefined distance from the user, we partially resolve the eye-camera parallax by warping the camera images through a perspective preserving homography.

The appropriate plane-induced homography between the image planes of the pair of stereo cameras and those of their associated displays, is computed by explicitly taking into consideration the geometry of the VST HMD and the intrinsic parameters of camera and display. The homography is therefore estimated uniquely by means of design and calibration data.

For validating our solution, we conducted objective and subjective tests. The goal of the tests was to assess the efficacy of such solution in recovering the natural perception of depth in the space around a pre-established distance from the observer.

Thanks to the hybrid nature of the HMD, which can work also under OST modality, in the first session of tests we measured the patterns of on-image disparities between a direct view of the world and a VST view. An internal camera, placed at the ideal eye's position, captured both the views of a target plane at different distances and orientations relative to the HMD.

These monoscopic disparities provided an initial measure of the amount of perceptual distortions brought by the non-orthoscopic setting of the HMD. The same disparities were then used to quantitatively estimate the resulting degree of uncertainty in perceiving relative depths under VST view.

Finally, we performed subjective tests aimed at assessing under real-use conditions, the actual depth estimation accuracy under VST view.

From a human-factor standpoint, VST HMDs raise issues related to the user's interaction with the augmented content and to some perceptual conflicts. With stereoscopic VST HMDs, the user can perceive relative depths between real and/or virtual objects by providing consistent depth cues in the recorded images delivered to the left and right eyes by the two displays of the visor. In our tests, we focused on relative depth measurements since relative depths information are much more important than absolute depths for aiding manual tasks in the personal (and intimate) space [7,8].

However, depth perception in binocular VST HMDs has not been fully investigated in literature. In their study, Kyto et al. [7] performed perceptual tests with a stereoscopic VST HMD aimed at measuring the effect of binocular disparities, relative size and height in the visual field on depth judgments in the action space (distances from 2 to 20 m). Their main finding was that depth perception through VST view in the action space is highly improved by a proper combination of a virtual content (i.e., auxiliary augmentations) providing binocular disparity and relative size cues. In our study, we did not use any sort of auxiliary augmentation since the goal of our perceptual studies was to assess how depth perception is recovered when using non-orthoscopic VST HMDs. Further, our depth judgment tests were performed within the personal and intimate space where the visual interaction with the augmented scene is likely to hide the ground plane and for which other depth cues other than binocular disparities and occlusions are not as relevant.

Overall, the obtained results were surprisingly positive in terms of depth judgment tasks. This is in line with what experienced by State et al. [19] and suggested by Milgram et al. [38], who both asserted that the distortion of the visual space derived from the mathematical models underpinning stereo vision is significantly higher than what the user perceives in reality. In our opinion, this fact is mostly motivated by the presence of other binocular depth cues as eyes convergence or monocular ones as linear perspective, texture gradient, shades and shadows [39]. All these cues contribute to provide a finer perception of depth in the personal space and partially compensate for the distortions brought by the non-orthoscopic setting of the VST HMD. The results of our preliminary perceptual tests were even more positive than the ones presented in [19], as in our tests the user could not use the hand as a "visual aid" for relative and absolute distance estimations.

Another aspect to consider is that the distortion of the patterns of binocular horizontal disparities at distances different from the homography plane, is not as severe at the centre of the stereo images, which is where the user normally directs his own view. Further, even if the vertical monoscopic disparities follow a steeper trend as the distance from the homography plane increases, we believe that their effect onto the perception of relative depths is not as evident, also considering that their combinatory effect is likely to be null.

In conclusion, the quasi-orthoscopic setting of the HMD and the user-specific homography, refined to embody the effects of the intrinsic parameters of the displays and of the non-ideal eyes placement in the HMD, are sufficient to recover a proper perception of relative depths in the personal space. Further, we can assert that the actual density of homographies that ensures a non-perceptible distortion of the visual space in the personal space can be sparser than the ideal pattern retrieved by estimating the trend of the stereoacuities of the HMD.

All of this suggests that we should investigate whether display calibration and eye-tracking can allow the achievement of similar results without the need for a user-specific homography refinement. Display calibration would in fact provide a proper estimation of the linear and non-linear projective parameters of the display, while eye-tracking would yield a robust and reliable estimation of the eyes position.

It is important to outline that the results of our perceptual tests can be considered as a preliminary proof of effectiveness of the proposed solution in recovering natural depth perception in a quasi-orthoscopic VST HMD. In addition, the testing platform herein used strongly encourages us to conducting structured user-studies involving more subjects and aimed at investigating further on how our solution can be of help to the VR and AR communities for investigations relative to user's perception and task achievement efficiency, hence in fields as human-computer interaction, neuroscience and human factor in computing systems.

Acknowledgments: This work was funded by the HORIZON2020 Project VOSTARS, Project ID: 731974. Call: ICT-29-2016—Photonics KET 2016.

Author Contributions: Fabrizio Cutolo, Umberto Fontana and Vincenzo Ferrari designed and developed the head-mounted display. Fabrizio Cutolo elaborated the proper scientific framework of the proposed solution and explained the differences between this paper and previous solutions in the field. Fabrizio Cutolo, Umberto Fontana and Vincenzo Ferrari conducted the tests and Fabrizio Cutolo and Umberto Fontana analyzed and discussed the results.

Conflicts of Interest: The authors declare no conflict of interest.

Appendix A

Hereafter the disparity maps for all the target planes considered are shown.

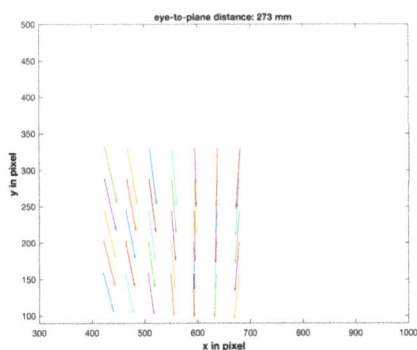

Figure A1. Map of disparities for a target plane placed at 273 mm from the observer.

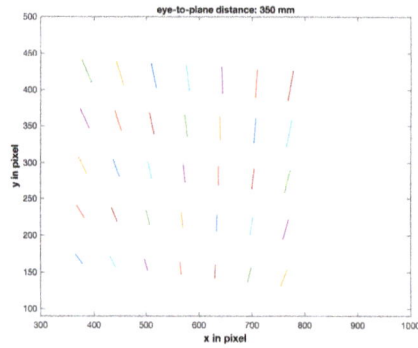

Figure A2. Map of disparities for a target plane placed at 350 mm from the observer.

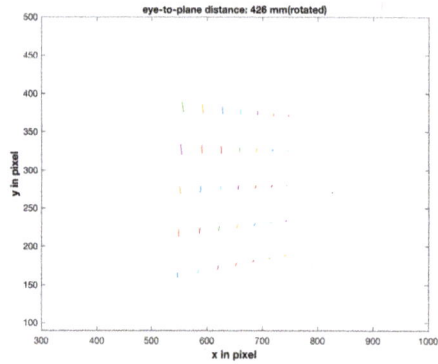

Figure A3. Map of disparities for a target plane rotated and placed at 426 mm from the observer.

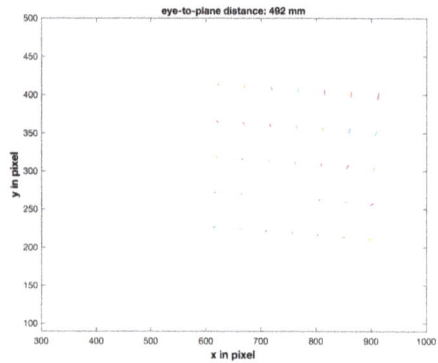

Figure A4. Map of disparities for a target plane placed at 492 mm from the observer.

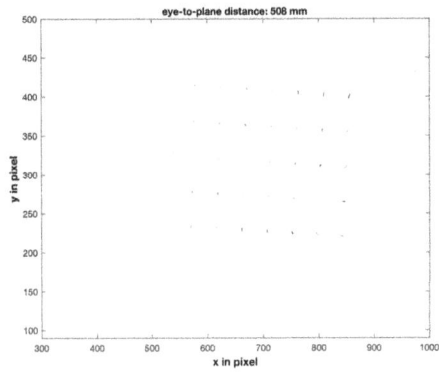

Figure A5. Map of disparities for a target plane placed at 508 mm from the observer.

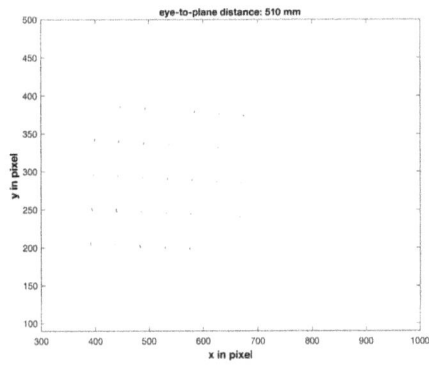

Figure A6. Map of disparities for a target plane placed at 510 mm from the observer.

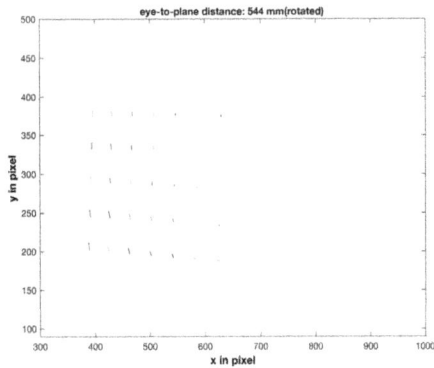

Figure A7. Map of disparities for a target plane rotated and placed at 544 mm from the observer.

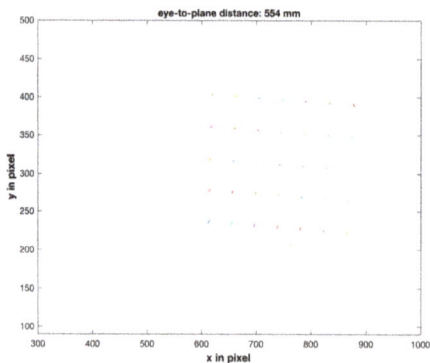

Figure A8. Map of disparities for a target plane placed at 554 mm from the observer.

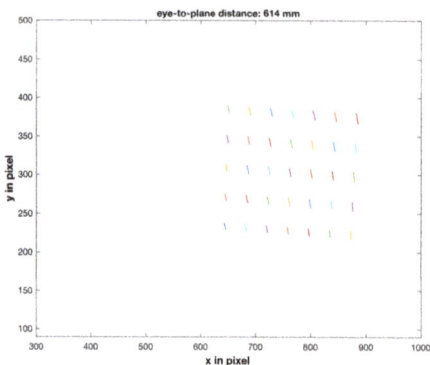

Figure A9. Map of disparities for a target plane placed at 614 mm from the observer.

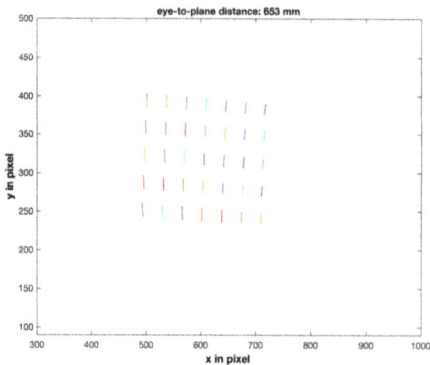

Figure A10. Map of disparities for a target plane placed at 653 mm from the observer.

References

1. Cutolo, F.; Freschi, C.; Mascioli, S.; Parchi, P.; Ferrari, M.; Ferrari, V. Robust and accurate algorithm for wearable stereoscopic augmented reality with three indistinguishable markers. *Electronics* **2016**, *5*, 59. [CrossRef]

2.   Rolland, J.P.; Fuchs, H. Optical versus video see-through mead-mounted displays in medical visualization. *Presence Teleoper. Virtual Environ.* **2000**, *9*, 287–309. [CrossRef]
3.   Benton, S.A. *Selected Papers on Three-Dimensional Displays*; SPIE Optical Engineering Press: Bellingham, WA, USA, 2001.
4.   Rolland, J.P.; Holloway, R.L.; Fuchs, H. A comparison of optical and video see-through head-mounted displays. *Telemanip. Telepresence Technol.* **1994**, *2351*, 293–307.
5.   Cutolo, F. Augmented reality in image-guided surgery. In *Encyclopedia of Computer Graphics and Games*; Lee, N., Ed.; Springer International Publishing: Cham, Switzerland, 2017; pp. 1–11.
6.   Kytö, M. *Depth Perception of Augmented and Natural Scenes through Stereoscopic Systems*; Aalto University: Espoo, Finland, 2014.
7.   Kyto, M.; Makinen, A.; Tossavainen, T.; Oittinen, P. Stereoscopic depth perception in video see-through augmented reality within action space. *J. Electron. Imaging* **2014**, *23*, 011006. [CrossRef]
8.   Cutting, J.E.; Vishton, P.M. Perceiving layout and knowing distances: The integration, relative potency, and contextual use of different information about depth. In *Perception of Space and Motion*; Academic Press: San Diego, CA, USA, 1995; pp. 69–117.
9.   Drascic, D.; Milgram, P. Perceptual issues in augmented reality. In Proceedings of the SPIE International Society for Optical Engineering, San Jose, CA, USA, 31 January–1 February 1996; pp. 123–134.
10.   Fuchs, H.; Livingston, M.A.; Raskar, R.; Colucci, D.; Keller, K.; State, A.; Crawford, J.R.; Rademacher, P.; Drake, S.H.; Meyer, A.A. Augmented reality visualization for laparoscopic surgery. In Proceedings of the First International Conference on Medical Image Computing and Computer-Assisted Intervention (MICCAI), Cambridge, MA, USA, 11–13 October 1998; pp. 934–943.
11.   Takagi, A.; Yamazaki, S.; Saito, Y.; Taniguchi, N. Development of a stereo video see-through hmd for ar systems. In Proceedings of the IEEE and ACM International Symposium on Augmented Reality, Munich, Germany, 5–6 October 2000; pp. 68–77.
12.   State, A.; Keller, K.P.; Fuchs, H. Simulation-based design and rapid prototyping of a parallax-free, orthoscopic video see-through head-mounted display. In Proceedings of the International Symposium on Mixed and Augmented Reality, Vienna, Austria, 5–8 October 2005; pp. 28–31.
13.   Bottecchia, S.; Cieutat, J.-M.; Merlo, C.; Jessel, J.-P. A new ar interaction paradigm for collaborative teleassistance system: The POA. *Int. J. Interact. Des. Manuf.* **2009**, *3*, 35–40. [CrossRef]
14.   Livingston, M.A.; Ai, Z.M.; Decker, J.W. A user study towards understanding stereo perception in head-worn augmented reality displays. In Proceedings of the 8th IEEE International Symposium on Mixed and Augmented Reality—Science and Technology, Orlando, FL, USA, 19–22 October 2009; pp. 53–56.
15.   Matsunaga, K.; Yamamoto, T.; Shidoji, K.; Matsuki, Y. The effect of the ratio difference of overlapped areas of stereoscopic images on each eye in a teleoperation. *Proc. Spiel. Int. Soc. Opt. Eng.* **2000**, *3957*, 236–243.
16.   Woods, A.; Docherty, T.; Koch, R. Image distortions in stereoscopic video systems. *Stereosc. Disp. Appl. IV* **1993**, *1915*, 36–48.
17.   Ferrari, V.; Cutolo, F.; Calabrò, E.M.; Ferrari, M. [Poster] HMD video see though AR with unfixed cameras vergence. In Proceedings of the IEEE International Symposium on Mixed and Augmented Reality (ISMAR), Munich, Germany, 10–12 September 2014; pp. 265–266.
18.   Cutolo, F.; Ferrari, V. The role of camera convergence in stereoscopic video see-through augmented reality displays. In Proceedings of the Future Technologies Conference (FTC), Vancouver, BC, Canada, 29–30 November 2017; pp. 295–300.
19.   State, A.; Ackerman, J.; Hirota, G.; Lee, J.; Fuchs, H. Dynamic virtual convergence for video see-through head-mounted displays: Maintaining maximum stereo overlap throughout a close-range work space. In Proceedings of the IEEE and ACM International Symposium on Augmented Reality, New York, NY, USA, 29–30 October 2001; pp. 137–146.
20.   Cutolo, F.; Fontana, U.; Carbone, M.; Amato, R.D.; Ferrari, V. [Poster] hybrid video/optical see-through HMD. In Proceedings of the 2017 IEEE International Symposium on Mixed and Augmented Reality (ISMAR-Adjunct), Nantes, France, 9–13 October 2017; pp. 52–57.
21.   LC-TEC Advanced Liquid Crystal Optics. Available online: http://www.lc-tec.se/ (accessed on 30 October 2017).
22.   Lumus. Available online: http://lumusvision.com (accessed on 30 October 2017).

23. Kanbara, M.; Okuma, T.; Takemura, H.; Yokoya, N. A stereoscopic video see-through augmented reality system based on real-time vision-based registration. In Proceedings of the IEEE Virtual Reality 2000 (Cat. No. 00CB37048), New Brunswick, NJ, USA, 18–22 March 2000; pp. 255–262.
24. Allison, R.S. Analysis of the influence of vertical disparities arising in toed-in stereoscopic cameras. *J. Imaging Sci. Technol.* **2007**, *51*, 317–327. [CrossRef]
25. Banks, M.S.; Read, J.C.A.; Allison, R.S.; Watt, S.J. Stereoscopy and the human visual system. *SMPTE Motion Imaging J.* **2012**, *121*, 24–43. [CrossRef] [PubMed]
26. Vienne, C.; Plantier, J.; Neveu, P.; Priot, A.E. The role of vertical disparity in distance and depth perception as revealed by different stereo-camera configurations. *I-Perception* **2016**, *7*. [CrossRef] [PubMed]
27. Zhang, Z.Y. A flexible new technique for camera calibration. *IEEE Trans. Pattern Anal.* **2000**, *22*, 1330–1334. [CrossRef]
28. Grubert, J.; Itoh, Y.; Moser, K.R.; Swan, J.E., II. A survey of calibration methods for optical see-through head-mounted displays. *IEEE Trans. Vis. Comput. Graph.* **2017**. [CrossRef] [PubMed]
29. Hartley, R.; Zisserman, A. *Multiple View Geometry in Computer Vision*; Cambridge University Press: Cambridge, UK, 2003.
30. Tomioka, M.; Ikeda, S.; Sato, K. Approximated user-perspective rendering in tablet-based augmented reality. In Proceedings of the 2013 IEEE International Symposium on Mixed and Augmented Reality (ISMAR)—Science and Technology, Adelaide, Australia, 1–4 October 2013; pp. 21–28.
31. Lothe, P.; Bourgeois, S.; Royer, E.; Dhome, M.; Naudet-Collette, S. Real-time vehicle global localisation with a single camera in dense urban areas: Exploitation of coarse 3D city models. In Proceedings of the 2010 IEEE Conference on Computer Vision and Pattern Recognition (CVPR), San Francisco, CA, USA, 13–18 June 2010; pp. 863–870.
32. Cutolo, F.; Siesto, M.; Mascioli, S.; Freschi, C.; Ferrari, M.; Ferrari, V. Configurable software framework for 2D/3D video see-through displays in medical applications. In *Augmented Reality, Virtual Reality, and Computer Graphics, Part II, Proceeding of the Third International Conference, AVR 2016, Lecce, Italy, 15–18 June 2016*; De Paolis, L.T., Mongelli, A., Eds.; Springer International Publishing: Cham, Switzerland, 2016; pp. 30–42.
33. Vtk Visualization Toolkit. Available online: https://www.vtk.org/ (accessed on 30 October 2017).
34. Opencv. Open Source Computer Vision Library. Available online: https://opencv.org/ (accessed on 30 October 2017).
35. Gadia, D.; Garipoli, G.; Bonanomi, C.; Albani, L.; Rizzi, A. Assessing stereo blindness and stereo acuity on digital displays. *Displays* **2014**, *35*, 206–212. [CrossRef]
36. Harris, J.M. Monocular zones in stereoscopic scenes: A useful source of information for human binocular vision? In *Stereoscopic Displays and Applications XXI*; SPIE: Bellingham, WA, USA, 2010; pp. 151–162.
37. Kyto, M.; Nuutinen, M.; Oittinen, P. Method for measuring stereo camera depth accuracy based on stereoscopic vision. In *Three-Dimensional Imaging, Interaction, and Measurement*; SPIE: Bellingham, WA, USA, 2011.
38. Milgram, P.; Kruger, M. Adaptation effects in stereo due to online changes in camera configuration. In *Stereoscopic Displays and Applications III*; SPIE: Bellingham, WA, USA, 1992; pp. 122–134.
39. Tovée, M.J. *An Introduction to the Visual System*; Cambridge University Press: Cambridge, UK, 1996.

technologies

MDPI

Article

A Low-Cost, Wearable Opto-Inertial 6-DOF Hand Pose Tracking System for VR

Andualem T. Maereg *, Emanuele L. Secco, Tayachew F. Agidew, David Reid and Atulya K. Nagar

Robotics Lab, Department of Mathematics and Computer Science, Liverpool Hope University, Liverpool L16 9JD, UK; seccoe@hope.ac.uk (E.L.S.); Agidewt@hope.ac.uk (T.F.A.); Reidd@hope.ac.uk (D.R.); nagara@hope.ac.uk (A.K.N.)
* Correspondence: maerega@hope.ac.uk; Tel.: +44-0741-745-0336

Received: 31 May 2017; Accepted: 25 July 2017; Published: 28 July 2017

Abstract: In this paper, a low cost, wearable six Degree of Freedom (6-DOF) hand pose tracking system is proposed for Virtual Reality applications. It is designed for use with an integrated hand exoskeleton system for kinesthetic haptic feedback. The tracking system consists of an Infrared (IR) based optical tracker with low cost mono-camera and inertial and magnetic measurement unit. Image processing is done on LabVIEW software to extract the 3-DOF position from two IR targets and Magdwick filter has been implemented on Mbed LPC1768 board to obtain orientation data. Six DOF hand tracking outputs filtered and synchronized on LabVIEW software are then sent to the Unity Virtual environment via User Datagram Protocol (UDP) stream. Experimental results show that this low cost and compact system has a comparable performance of minimal Jitter with position and orientation Root Mean Square Error (RMSE) of less than 0.2 mm and 0.15 degrees, respectively. Total Latency of the system is also less than 40 ms.

Keywords: optical tracking; inertial tracking; sensor fusion; virtual reality; low cost

1. Introduction

Physical immersion and highly interactive systems are important for effective virtual reality applications. User interactions in Virtual Reality (VR) can be displayed in the form of visual, aural and haptic sensory modalities [1–3]. Continuous hand tracking is crucial for a more realistic and immersive virtual experiences. Commercial VR devices such as Oculus Rift headset and HTC VIVE™ with the integration of hand tracking systems like Leap Motion Controller enables us to experience "visually realistic" interaction with Virtual objects. However, most of these commercial devices does not provide touch feedback (haptics). The integration of haptics in VR devices will improve interactivity and immersion [4]. Fully optical devices like Leap Motion have limited applicability for VR haptic devices. The main reason being the haptic setup on the hand can occlude part of the skin which affects the performance of the tracker. This motivates the development of a low cost hand tracking system for an integration with a lightweight, low cost, wearable and wireless exoskeleton setup for force feedback.

A variety of hand tracking methods have been developed for different application areas including VR. These mechanisms are mainly based on optical, inertial, mechanical, electromagnetic and acoustic sensors [5]. The main consideration in the choice of tracking systems are accuracy and precision, update rate of tracking outputs, robustness to interference and occlusions and encumbrance from wires and mechanical linkages. The presence of a hand exoskeleton in a haptic setup would make use of the mechanical tracking system an easy approach. Mechanical tracking problems involve estimating the motion of one link relative to the attached moving link [6]. Hence, hand tracking problems can be solved by estimating the position of each bone relative to the previous bone. In this case, the entire hand can be arranged as a sequence of attached rigid bodies. The position of the bodies relative to each other can also be solved using multibody kinematics which also enables us to determine the velocity and acceleration of attached bodies. The known kinematic constraints of the hand model improve the accuracy of tracking [7]. This kind of model based hand tracking can be more effective for applications which need hand exoskeletons for Haptics and Rehabilitation [8]. Although the position and orientation of each bone on the hand can be estimated accurately with the hand exoskeleton, tracking of reference frame on the palm or wrist is necessary. Palm position and orientation is used to estimate the pose of the overall hand based on a predefined kinematic hand model.

In this paper, a low cost six Degree of Freedom (6-DOF) tracking of the single point reference frame on palm is discussed using opto-inertial approach. This system aims at providing a relative pose estimation of human hand while immersed in VR environment and wearing an integrated exoskeleton for full hand configuration estimation. Therefore, the current design of the proposed tracking system is part of a mechanical exoskeleton tracking and force feedback system. The system offers a good compromise in terms of cost and efficiency for VR applications with a predefined workspace range and also haptic setups which need hand tracking as an input. In the proposed tracking system, the position and orientation estimation are handled separately. Position is estimated from optical tracking outputs while orientation is estimated from the Inertial Measurement Unit (IMU) tracking outputs. Since we adopt a geometrical approach to estimate the depth tracking, orientation data has been used to correct estimation error. In addition, yaw angle estimation involves the use of optical Infrared (IR) position outputs as a complementary system to reduce IMU heading error from magnetic interference on yaw calculation.

Various optical systems have been designed, typically using video cameras and several Infrared Light Emitting Diodes (LEDs) [9]. Optical systems do not require mapping and provide relatively high accuracies over a large workspace. However, a constant line of sight between the IR LEDs and the camera must be maintained [10]. The performance of this optical system can be affected by occlusion and limiting the coverage area. To overcome the limitations in camera view range, pan-tilt-zoom cameras can be used with visual servoing techniques which follows the markers to be tracked [11]. It is also difficult to insure a proper tracking under different lighting conditions. Inertial sensors, on the other hand, have no range limitation and no line of sight is required. They can give high bandwidth motion measurement with negligible latency. However, they are prone to an interference from magnetic fields [5]. Due to these complementary pros and cons of the two tracking systems, combination of optical and inertial technologies results in more accurate 6-DOF pose tracking [12,13].

Optical Tracking systems detect and track artificial features such as normal LEDs and IR LEDS (active features) or retroactive materials illuminated by Infrared light and a special tag placed onto the hand (passive features). Using passive features could be a problem in the absence of sufficient light; therefore, using active features which emits light is more reliable [13]. To reduce the problems of optical motion tracking related to lightning, Infrared LEDs can be used to completely isolate the markers from the background light. However, this comes at the expense of requiring a special IR camera. Most low cost IR tracking systems use normal Webcams with some modification to pass IR light. Normal Webcams have an infrared blocking filter which prevent the IR from entering. This can be addressed by removing the filter, so that the camera can be sensitive to infrared light. IR detection is more reliable in all lighting conditions; however, for a more accurate tracking a infrared pass-filter can also be added blocking most of the visible light spectrum.

Even though accuracy and precision have lesser importance for VR environments unlike tracking for localization [14], resolution of less than 1 mm and angular precision of greater than 0.2 degrees is important for VR applications. Tracking latency beyond 40–60 ms will also affect the performance of the VR [15].

2. Materials and Methods

The optical tracking part consists of PlayStation 3 (PS3) eye camera (costs £8) and two IR LEDs while the inertial tracker consists of an LSM9DS0 IMU module which contains a 3-axis gyroscope, 3-axis accelerometer and 3-axis magnetometer in a single chip. The IMU provides a 9-DOFs data stream to the Mbed LPC1768 Microcontroller (32-bit ARM Cortex-M3 running at 96 MHz). A LabVIEW software captures orientation data via Bluetooth at rate of 60 Hz. LabVIEW timed loop enables to trigger reading of both optical and IMU outputs with similar timing sources thereby synchronizing the IR tracking position outputs with the orientation outputs and transfer the 6-DOF pose of the palm through User Datagram Protocol (UDP) connection stream to virtual environment made in Unity 3D.

2.1. Optical Tracking System

Optical Tracking System relies in estimating the pose of the hand from 3D features or targets attached on the hand. In this experiment, two IR LEDs are used as targets. The IR tracking system uses a blob detection and tracking image processing algorithm for positional tracking in 3-DOF. The camera is placed above the workspace (as shown in Figure 1) and tracks 2 IR LEDs placed on the setup. The position of the camera is selected purposely to eliminate the effect of direct light through doors, windows and other light sources. One or two IR LEDs enables us to track 2-DOF while 3 or more LEDs can track 3-DOF position and 3-DOF orientations. The LED viewing angle is also important. Most LEDs focus the light narrowly; therefore, to distribute the light evenly in all direction a diffuser foam cover is used. Wide angle LED can also be used.

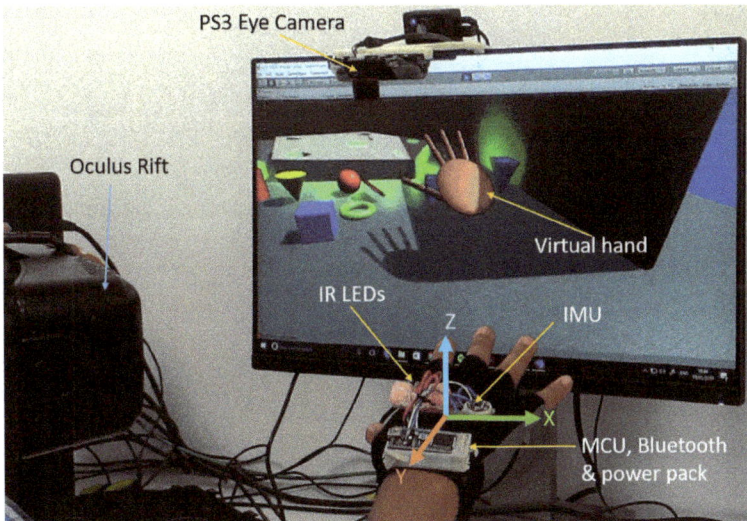

Figure 1. Virtual reality 6 Degree of Freedom (6-DOF) tracking experimental setup.

The computer vision algorithms are implemented using LabVIEW National Instrument (NI) vision algorithms. Image frames are acquired continuously with PS3 Eye Camera and NI image acquisition software is used for inline processing. Image acquisition setting are configured as video resolution of 320 × 240 at 60 fps. The infrared LEDs are tracked using a simple image processing algorithm called blob tracking. Blob tracking is more reliable and effective than most complicated image tracking algorithms in a wide range of lighting environment and it can also be done at full frame rate with minimum Central Processing Unit (CPU) usage.The IR LEDs are detected as an area of high brightness in the image. Considering the camera resolution, view range and angle, the workspace of the tracking system is limited to the range of 40 cm width, 30 cm length and 30 cm height. Segmentation is used to distinguish regions (set of pixels), which corresponds to IR markers and the background. RGB ranges are used as a criterion to decide whether a pixel belongs to a region of interest or background. Thresholding applies a threshold of the RGB image in range of Red (0–255), Green (0–255) and Blue

(0–202). Pixels outside this range are selected as a region of interest and all other pixels are classified as background.

Thresholding results in an 8 bit grayscale image, which has to be inverted to reverse the dynamic of the image. Different morphology techniques are then applied to remove small blobs, and fill holes in the large blobs detected. These morphological operators are selected among other possible operators because they do not affect the estimation of the coordinates of the centroid of the blobs. As shown in Figure 2, the final blobs are characterized by a smooth profile with very much contained glares. A bounding circle is formed around the blobs which helps to extract the measurement results such as number of blobs detected, centre of mass X and centre of mass Y of the blobs. The centre of mass measurements in the X and Y directions given by image coordinates (u, v) are mapped into x and y world coordinates. Figure 3 shows the main blocks of the vision algorithms used to track the IR markers.

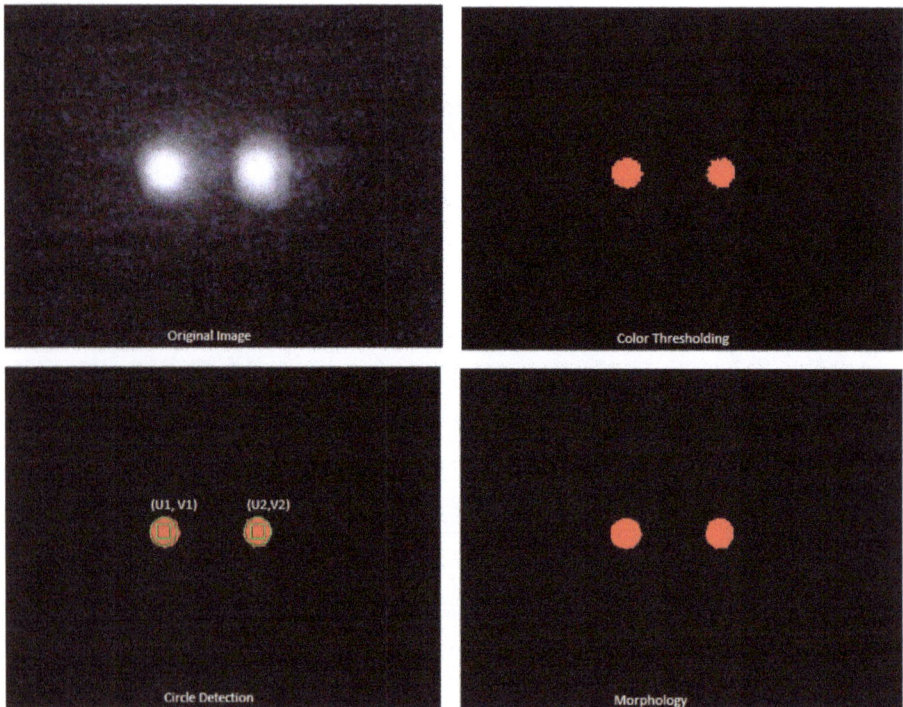

Figure 2. Blob detection and tracking algorithm outputs.

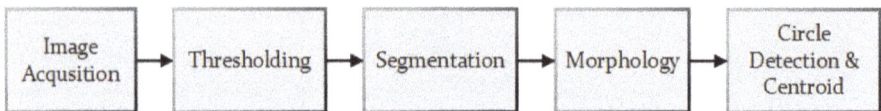

Figure 3. Blob tracking: Image processing algorithms.

A moving body can be tracked using n observed blobs. Each blob corresponds to a point p (x, y, z) with coordinates defined with respect to a reference frame in world coordinate system [16]. The point observed at image coordinate (u, v) provides information about 2-DOFs. If both IR LED are in the field of view of the camera, considering movement on 2D plane only, the x and y position can be estimated from the image coordinates detected as in the following equations.

$$x_m = C_x \frac{u_1 + u_2}{2}, y_m = C_y \frac{v_1 + v_2}{2} \tag{1}$$

Where C_x and C_y are scaling factors from image coordinate to real world coordinates. However, For 3-DOF positional tracking, the distance from the camera to the features has to be known. Many algorithms have been developed to determine the depth by solving point model problems with 3 or more IR spots, multiple cameras or stereo cameras can also be used. Using multiple or stereo camera systems increases the complexity and cost of the system where as 3 or more IR targets tracking creates non-stable tracking outputs [17]. In VR applications, considerable spatial accuracy errors are more acceptable than drifts and jitter since users continuously use visual feedback to correct positional errors. On the other hand, drift and jitter creates discomfort for the user. As result, an easier approach has been used to map the depth considering the trade-off between accuracy, cost and computational complexity. This approach is based on mapping the depth with the proportional distance between the 2 IR LEDs.

$$d_{measured} = C_z \left((v_2 - v_1)^2 + (u_2 - u_1)^2 \right)^{\frac{1}{2}} \tag{2}$$

It is clear that the projection of the distance between the two IR LEDs varies with rotation of the hand with respect to the y axis (roll angle); therefore, the depth calculation can be corrected by adding the roll angle (θ) as factor.

$$z_m = \frac{d_{measured}}{\cos \theta} \tag{3}$$

A clear z axis only motion of the IR spots can also produce an error in the x and y axis. This error increases as the position of IR spots is far from the optical axis of the camera. However, the error increases linearly as distance increase from origin; therefore, a calibration matrix (mapping matrix) can be formulated to reduce the error in the x and y axis reading caused by z axis movements. An automatic linear fitting method is used to find the relationship between the depth values and the change in X and Y reading caused by Z axis movement. The following steps have been implemented to find the slope of the deviation in the X and Y axis.

1. Find the linear fitting values from the graph of Z vs. X axis and Z vs. Y axis while moving the IR tracker in the Z axis. Slope and intercept is calculated for 10 sample points. To calculate slope and intercepts of data sequence (X, Y)using a least square solution, the LabVIEW linear fit Virtual Instrument (VI) uses the iterative general Least Square method to fit points to a straight line of the form

$$f = mx + b \tag{4}$$

where x is the data sequence, m is slope and b is intercept. Every iteration gives linear curve of the form

$$y_i = mx_i + b \tag{5}$$

The least square method finds the slope and intercept which minimizes the residue expressed by the following equation.

$$\frac{1}{N} \sum_{i=0}^{N-1} (f_i - y_i)^2 \tag{6}$$

where N is the length of Y, fi is the ith element of Best Linear Fit, and yi is the ith element of Y.

2. Feedback and update the new slope values as coefficient of the calibration matrix (m_x and m_y)
3. Continue to Step 1

Reducing the x and y positional components caused by z axis movements from the pure x and y position reading avoids the error.

$$\begin{bmatrix} x_{cal} \\ y_{cal} \\ z_{cal} \end{bmatrix} = \begin{bmatrix} 1 & 0 & -m_x \\ 0 & 1 & -m_y \\ 0 & 0 & 1 \end{bmatrix} \begin{bmatrix} x_m \\ y_m \\ z_m \end{bmatrix} \tag{7}$$

This calibrated 3-DOF position can give us smooth and proportional movement mapping between the real hand and virtual hand. Figure 4 shows a test for the above algorithm for a square trajectories on the X-Z plane.

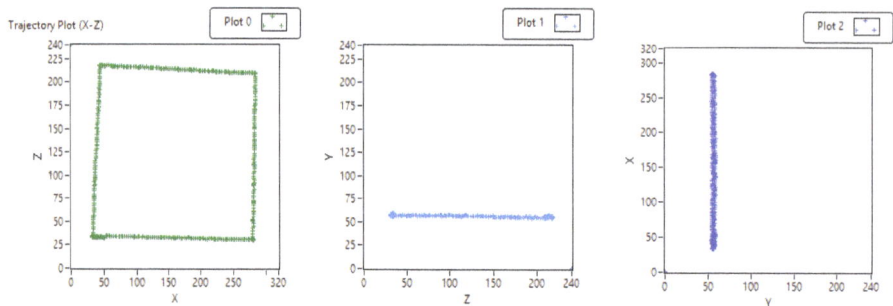

Figure 4. Square trajectory plot: Calibration tests in X-Z plane, Z-Y plane and Y-X plane.

2.2. Inertial Tracking System

The 9-DOF LSM9DS0 MARG (Magnetic, Angular Rate and Gravity) sensor is used to obtain highly accurate orientation tracking with high update rate. The gyroscope measures angular velocity along the three orthogonal axis, accelerometers measure linear acceleration and magnetometer measure magnetic field strength along the three perpendicular axis providing an absolute reference of orientation.

Kalman filters are the most widely used orientation filter algorithms. However, they are complicated for implementation and demand a large computational load which makes it difficult for implementation on small scale microcontrollers. The Magdwick filter has been used as an alternative approach. This filter is effective at low sampling rates, and is more accurate than the Kalman-based algorithm and has low computational load. The MARG system also known as AHRS (Altitude and Heading Reference Systems) is able to provide a measurement of orientation relative to the earth magnetic field and direction of gravity. The algorithm uses a quaternion representation of orientation to reduce singularities associated with Euler angle representations [18].

The magwick filter includes an online gyroscope bias drift compensation. The gyroscope zero bias drift overtime caused by temperature and motion with time. Mahony et al. [19] showed that the gyroscope bias drift can also be compensated by orientation filter through integral feedback in the rate of change of orientation. Magdwick approaches have also implemented a similar algorithm for gyroscope drift compensation [18].

Pitch, roll and yaw angles can be purely estimated from quaternion values. However, estimation of yaw angle involves magnetometer data which can be affected by elevation and tilt angle as well as hard and soft iron bias. Hard iron biases can be removed using different calibration techniques [20]. Soft iron biases cause errors in the measured direction of the Earth's magnetic field. Declination errors need additional reference heading while inclination errors can be compensated for using the accelerometer as it provides an additional measurement of the sensor's attitude. The magnetometer used in our application is calibrated to reduce the yaw drift error. However, due to a nearby magnetic object, the sensor reading becomes more dependent on the change in place. Such cases reduce the reliability of the magnetometer to produce an accurate yaw orientation. Therefore, the camera information is used as an additional source of detecting yaw orientation. Since the camera optical axis is parallel to the yaw axis, two IR LEDs position information can be used to estimate the yaw orientation easily.

$$yaw = \arctan 2 \left(\frac{v_2 - v_1}{u_2 - u_1} \right) \tag{8}$$

2.3. Performance Evaluation

Tests has been done to characterize the resolution, static and dynamic spatial accuracy of the overall tracking system. The resolution is the smallest change of position or orientation that the tracking system is able to detect. Resolution is limited either by jitter or quantization levels. The visual effect of the jitter on computer display can affect the user's haptic experiences. Static accuracy is the amount of reading error when the position and orientation remain constant. Errors due to noise, scale factor error and non-linearity can be shown on static accuracy tests. Very low frequency error components which can be perceived over a period of time are categorized under the term stationary drift. The jitter is the rapidly changing error component [5]. The static accuracy is calculated as the Root Mean Square (RMS) error of the recorded position and orientation angles from the true ones when the tracking sensors are held at a known fixed position and recording the position and orientation output data stream for a 10 min period of time.

For the position static accuracy test, the target LEDs are taped on a sheet of paper with a square grid. The grid origin is aligned with the camera optical axis. A variety of known fixed positions and orientations are tested to see if the static accuracy could vary significantly depending on the position. From the static experimental results, drift and jitter can be seen due to the quantization error.

In order to remove jitter and drifts, smoothing filters of rectangular moving average are used. The moving average filter is an optimal filter to reduce random noises and retain sharp response. It is especially used for time domain signals. The moving average filter operates by averaging a number of points from the input signal to produce each point in the output signal. In our case, all samples in the moving average window are weighted equally to remove spikes in the signal. The moving average can be expressed in the equation form as

$$y[i] = \frac{1}{M} \sum_{j=0}^{M-1} x[i+j] \tag{9}$$

Where $x[i]$ is the input signal, $y[i]$ is the output signal, and M is the number of points on average. The result shows that most of the jitter and drift is voided by using the filter. Removing the continuous jitter is particularly important for VR application, even more so than having spatial accuracy. For angular static accuracy testing, we mounted the IMU on Goniometer attached with a stable box. Orientation outputs are recorded for fixed roll, pitch and yaw angles.

Dynamic positional accuracy tests are done with a calibrated 3D printer head which can be manually driven with a resolution of 10 mm or 20 mm jogging mode in the x and y axis and 10 mm jogging mode in the z axis as a "ground truth" reference system. Dynamic position readings are recorded while moving the 3D printer head in the x, y and z position. Angular dynamic accuracy tests are performed on the two axis gimbals shown in Figure 5. Angular data streams are recorded while rotating the gimbals with a mounted servo motor.

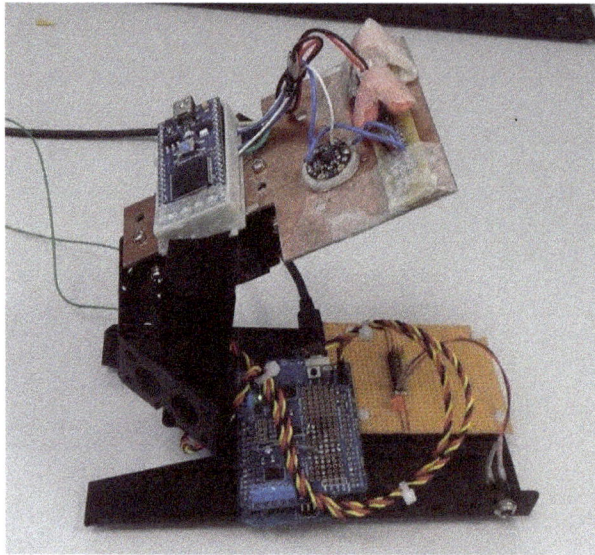

Figure 5. Orientation dynamic accuracy test setup.

3. Results

Position Data has been collected by placing the IR targets at fixed 3D positions on square grid for 10 min. The recorded data is used to compute the Root Mean Square Error (RMSE) by subtracting the mean value from each set of position readings according to the following equation.

$$RMSE = \sqrt{\frac{\sum_{t=1}^{n} (y_t - \overline{y_t})^2}{n}} \tag{10}$$

Where y_t are recorded data values and $\overline{y_t}$ is the mean value of n data points. The resulting RMSE for x, y and z are 0.288 mm, 0.268 mm and 0.653 mm respectively. Most of these errors are caused by quantization levels. This is because the position readings are directly mapped with the position of the blob centres in pixels which are always integer values. A camera with high resolution can give us a more accurate reading. The jitter is reduced by filtering to avoid the rapid flickering of the virtual hand. As shown in Table 1 the position RMSE error is reduced using a rectangular moving average smoothing filter to 0.148 mm, 0.104 mm and 0.373 mm in x, y and z direction respectively.

Table 1. Position and orientation Root Mean Square Error (RMSE): Static accuracy test.

	Position RMSE (mm)		Orientation RMSE (Degrees)		
Axis	Original	Filtered	Angle	Original	Filtered
X	0.288	0.148	Pitch	0.199	0.113
Y	0.268	0.104	Roll	0.137	0.079
Z	0.653	0.373	Yaw	0.831	0.486

To test the dynamic performance of the IR tracker, data are collected while moving the IR targets mounted on a 3D printer head. The printer head moves for a continuous 10 mm jog mode in one direction and repeats the same movement in the opposite direction where each 10 mm movement has constant velocities.

For static orientation tests, the IMU is fixed at different roll, pitch and yaw angles measured by Goniometer. Test results show that orientation results are accurate except for some drift, which occurred due to sensor bias or noise. To reduce the drift, the same smoothing filter has been used and RMSE error are reduced to 0.113°, 0.709° and 0.486° pitch, roll and yaw angle, respectively. Static position and orientation rest results are shown in Figure 6.

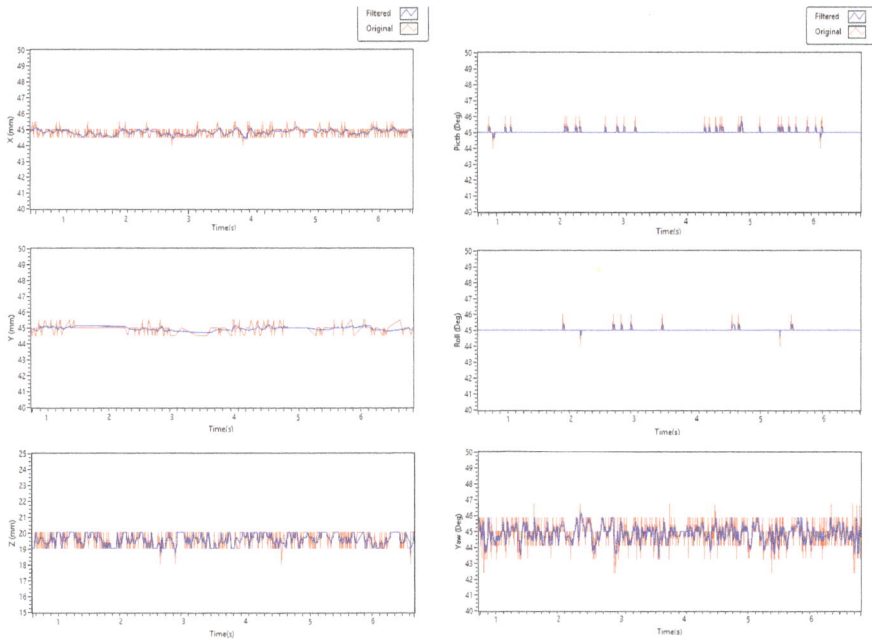

Figure 6. Static position and orientation accuracy test results.

To validate dynamic orientation accuracy, data has been collected while moving the IMU placed on top of the gimbal setup shown in Figure 5. The servos move the gimbals clockwise and counter clockwise repeatedly with constant speed. The speed is reduced to match the update rate of the IMU. Results (as shown in Figure 7) indicate that dynamic orientation is worse at higher speed movement, which may not occur on virtual interactions. Some of the error can also be caused by the jitter of the servos while moving. On the other hand, yaw orientation dynamic tests are free from such errors since optical tracking is mainly used to get the yaw angle except during occlusion of the IR targets.

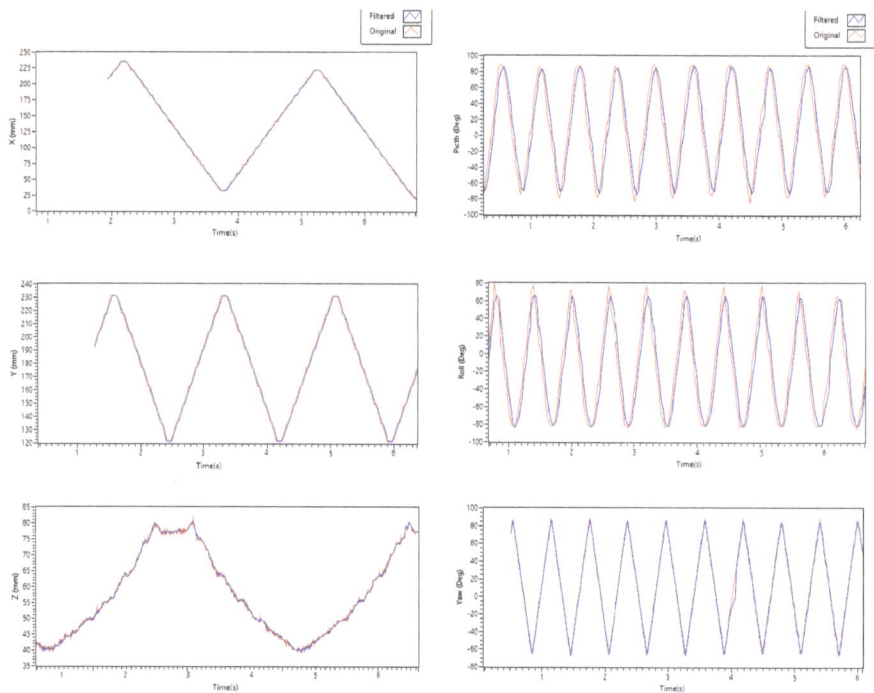

Figure 7. Position and orientation dynamic accuracy tests.

4. Conclusions

This paper presents a low cost and wearable approach for integration of hand tracking systems in a VR environment. Even though more accurate systems exist on the market and also as research equipment, none of them are affordable and they cannot be easily integrated in a VR environment with haptic setups. Accurate tracking systems which are based on fully optical technologies also have limitations in the presence of haptic devices due to occlusion. These limitations motivate us for the development of tracking systems which can be wearable and easily integrated with VR and haptics applications. Experimental Results shows that our approach can estimate the 6-DOF pose of the hand with reasonable accuracy and low latency. Future works will include a full haptic exoskeleton system which tracks full finger joints positions for a more immersive haptic experience.

Acknowledgments: The authors would like to acknowledge Liverpool Hope University for the research funding and covering the cost of this publication.

Author Contributions: A.T.M., E.L.S. and D.R. conceived and designed the experiment; A.T.M. and T.F.A. develop the experimental setup and perform the experiments; A.T.M. and E.L.S. analyzed the data; D.R. and A.K.N. contributed analysis tools; A.T.M. wrote the paper.

Conflicts of Interest: The authors declare no conflict of interest.

References

1. Maereg, A.T.; Secco, E.L.; Agidew, T.F.; Diaz-Nieto, R.; Nagar, A. Wearable haptics for VR stiffness discrimination. In Proceedings of the European Robotics Forum, Edinburgh, UK, 22–24 March 2017.
2. Andualem, T.M.; David, R.; Atulya, N.; Emanuele, L.S. Integrated wireless and wearable haptics system for virtual interaction. In Proceedings of the EuroHaptics, London, UK, 4–7 July 2016.

3. Li, M.; Konstantinova, J.; Secco, E.L.; Jiang, A.; Liu, H.; Nanayakkara, T.; Seneviratne, L.D.; Dasgupta, P.; Althoefer, K.; Wurdemann, H.A. Using visual cues to enhance haptic feedback for palpation on virtual model of soft tissue. *Med. Biol. Eng. Comput.* **2015**, *53*, 1177–1186.
4. Margolis, T.; DeFanti, T.A.; Dawe, G.; Prudhomme, A.; Schulze, J.P.; Cutchin, S. Low cost heads-up virtual reality (HUVR) with optical tracking and haptic feedback. In Proceedings of the Society of Photo-optical Instrumentation Engineers (SPIE), San Francisco, CA, USA, 23–27 January 2011.
5. Foxlin, E.; Altshuler, Y.; Naimark, L.; Harrington, M. Flighttracker: A novel optical/inertial tracker for cockpit enhanced vision. In Proceedings of the 3rd IEEE/ACM International Symposium on Mixed and Augmented Reality, Washington, DC, USA, 2–5 November 2004.
6. Gu, X.; Zhang, Y.; Sun, W.; Bian, Y.; Zhou, D.; Kristensson, P.O. Dexmo: An Inexpensive and Lightweight Mechanical Exoskeleton for Motion Capture and Force Feedback in VR. In Proceedings of the CHI Conference on Human Factors in Computing Systems, Santa Clara, CA, USA, 7–12 May 2016.
7. Steven, M.L. *Virtual Reality*; Cambridge University Press: Cambridge, UK, 2016.
8. Secco, E.L.; Sottile, R.; Davalli, A.; Calori, L.; Cappello, A.; Chiari, L. VR-Wheel: A rehabilitation platform for motor recovery. In Proceedings of the Virtual Rehabilitation, Venice, Italy, 27–29 September 2007.
9. Zaoui, M.; Wormell, D.; Altshuler, Y.; Foxlin, E.; McIntyre, J. A 6 DOF opto-inertial tracker for virtual reality experiments in microgravity. *Acta Astronaut.* **2001**, *49*, 451–462.
10. He, C.; Kazanzides, P.; Sen, H.T.; Kim, S.; Liu, Y. An inertial and optical sensor fusion approach for six degree-of-freedom pose estimation. *Sensors* **2015**, *15*, 16448–16465.
11. Cortes, G.; Marchand, É.; Ardouinz, J.; Lécuyer, A. Increasing optical tracking workspace of VR applications using controlled cameras. In Proceedings of the IEEE Symposium on 3D User Interfaces (3DUI), Los Angeles, CA, USA , 18–19 March 2017.
12. Hogue, A.; Jenkin, M.; Allison, R.S. An optical-inertial tracking system for fully-enclosed VR displays. In Proceedings of the First Canadian Conference on Computer and Robot Vision, London, ON, Canada, 17–19 May 2004.
13. Patel, K.; Stuerzlinger, W. Simulation of a virtual reality tracking system. In Proceedings of the IEEE International Conference on Virtual Environments Human-Computer Interfaces and Measurement Systems (VECIMS), Ottawa, ON, Canada, 19–21 September 2011.
14. Calloway, T.; Megherbi, D.B. Using 6 DOF vision-inertial tracking to evaluate and improve low cost depth sensor based SLAM. In Proceedings of the IEEE International Conference on Computational Intelligence and Virtual Environments for Measurement Systems and Applications (CIVEMSA), Budapest, Hungary, 27–28 June 2016.
15. Pintaric, T.; Kaufmann, H. Affordable infrared-optical pose-tracking for virtual and augmented reality. In Proceedings of the IEEE VR Workshop on Trends and Issues in Tracking for Virtual Environments, Charlotte, NC, USA, 11 March 2007.
16. Marchand, E.; Uchiyama, H.; Spindler, F. Pose estimation for augmented reality: A hands-on survey. *IEEE Trans. Vis. Comput. Graph.* **2016**, *22*, 2633–2651.
17. Satyavolu, S.; Bruder, G.; Willemsen, P.; Steinicke, F. Analysis of IR-based virtual reality tracking using multiple Kinects. In Proceedings of the IEEE Virtual Reality Short Papers and Posters (VRW), Costa Mesa, CA, USA, 4–8 March 2012.
18. Madgwick, S.O.; Harrison, A.J.; Vaidyanathan, R. Estimation of IMU and MARG orientation using a gradient descent algorithm. In Proceedings of the IEEE International Conference on Rehabilitation Robotics (ICORR), Zurich, Switzerland, 29 June–1 July 2011.
19. Mahony, R.; Hamel, T.; Pflimlin, J.M. Nonlinear complementary filters on the special orthogonal group. *IEEE Trans. Autom. Contr.* **2008**, *53*, 1203–1218.
20. Vasconcelos, J.F.; Elkaim, G.; Silvestre, C.; Oliveira, P.; Cardeira, B. Geometric approach to strapdown magnetometer calibration in sensor frame. *IEEE Trans. Aerosp. Electron. Syst.* **2011**, *47*, 1293–1306.

technologies

MDPI

Article

Antepartum Fetal Monitoring through a Wearable System and a Mobile Application

Maria G. Signorini [1], **Giordano Lanzola** [2,3], **Emanuele Torti** [2,3], **Andrea Fanelli** [1,4] and **Giovanni Magenes** [2,3,*]

[1]   Department of Electronics, Information and Bioengineering—DEIB, Politecnico di Milano,
     20133 Milano, Italy; mariagabriella.signorini@polimi.it (M.G.S.); fanelli@mit.edu (A.F.)
[2]   Department of Electrical, Computer and Biomedical Engineering, University of Pavia, 27100 Pavia, Italy;
     giordano.lanzola@unipv.it (G.L.); emanuele.torti@unipv.it (E.T.)
[3]   Center for Health Technologies (CHT), University of Pavia, 27100 Pavia, Italy
[4]   Integrative Neuromonitoring and Critical Care Informatics Group, Massachusetts Institute of Technology,
     Cambridge, MA 02139, USA
*   Correspondence: giovanni.magenes@unipv.it; Tel.: +39-35-644-2047

Received: 11 February 2018; Accepted: 24 April 2018; Published: 26 April 2018

Abstract: Prenatal monitoring of Fetal Heart Rate (FHR) is crucial for the prevention of fetal pathologies and unfavorable deliveries. However, the most commonly used Cardiotocographic exam can be performed only in hospital-like structures and requires the supervision of expert personnel. For this reason, a wearable system able to continuously monitor FHR would be a noticeable step towards a personalized and remote pregnancy care. Thanks to textile electrodes, miniaturized electronics, and smart devices like smartphones and tablets, we developed a wearable integrated system for everyday fetal monitoring during the last weeks of pregnancy. Pregnant women at home can use it without the need for any external support by clinicians. The transmission of FHR to a specialized medical center allows its remote analysis, exploiting advanced algorithms running on high-performance hardware able to obtain the best classification of the fetal condition. The system has been tested on a limited set of pregnant women whose fetal electrocardiogram recordings were acquired and classified, yielding an overall score for both accuracy and sensitivity over 90%. This novel approach can open a new perspective on the continuous monitoring of fetus development by enhancing the performance of regular examinations, making treatments really personalized, and reducing hospitalization or ambulatory visits.

Keywords: tele-monitoring; wearable devices; fetal heart rate; telemedicine

---

1. Introduction

In the last two decades, research has been very active in developing wearable devices and systems for medical-oriented applications, often aimed at continuously monitoring patients in their environment and during daily life activities. The need is a sustainable health system that can manage acute care (in hospital or emergency departments), and care of outpatients (mainly chronic), including healthy citizens during their normal life, in order to prevent possible diseases by means of tele-monitoring [1].

Until the most recent developments in Information and Communication Technologies (ICT), some limitations in connections and transmission rates allowed only collecting and sending few parameters in real-time, but nowadays it is possible to realize continuous surveillance of physiological signals. Thanks to the introduction of new miniaturized sensors and devices, together with the improvements in wireless transmission, the remote management of chronic diseases has become a reality, and studies prove that patients are willing to adopt it [2].

Nowadays, the switch to domiciliary care for non-critical medical issues is seen in many health care areas as a way to achieve better control over chronic diseases and consequently delay the occurrence of any complications, avoid unneeded hospitalizations, and reduce pressure over national budgets [3]. Many exploitations of structured tele-monitoring systems have already been carried out in patients with heart failure [4,5]. The CarelinkTM Network, for example, has been introduced by Medtronic to improve the management of patients suffering from heart failure that undergo treatments based on Implantable Cardioverter Devices (ICD) [6]. The literature shows that whenever ICDs are coupled with a remote monitoring service, not only it is possible to earlier detect any problem concerning the devices themselves, but also the therapy may be better optimized and individualized for each patient [7]. Even research projects investigating new treatments that pose significant threats to outpatients' health are increasingly relying on a real-time telemonitoring of clinical parameters for immediate processing and early detection of abnormal or hazardous conditions [8].

Furthermore, the great comfort of subjects in wearing miniaturized sensors, along with an increasing general health care policy of prevention, have promoted remote monitoring applications also in the healthy population. Recently, wearable systems have been adopted to monitor healthy subjects acting in extreme environments [9], performing highly risky activities such as firefighting [10], doing physical training [11], or wellness exercises [12]. A boost in this direction is given by the appearance of new textile materials that allow embedding low-cost wearable sensors and computing devices in standard duty outfits. Sensorized garments combined with tele-monitoring services may therefore be applied also to healthy people for detecting signals in subjects while they are engaged in their daily activities with the aim of preventing or anticipating the occurrence of unusual or pathological conditions.

Pregnancy and fetal states may also effectively resort to tele-monitoring. Although pregnancy in itself is not a pathology, the fetal condition should be monitored as often as possible in order to prevent unfavorable outcomes. The literature shows that pregnant women could highly benefit by a remote monitoring service that is particularly effective for those patients and also helps in achieving a more efficient use of the health care [13,14]. This is because the reduction of in- or out-stays for patients only undergoing routine examinations mitigates the shortages of beds in hospitals and reduces any related management costs. Therefore, it is desirable to design and implement wearable systems that allow the effective monitoring of pregnant women in different life conditions and evaluate both possible acute effects (emergency, risky conditions for the fetus), or non-acute effects related to unhealthy lifestyles. These factors, though not immediately life-threatening, may be the cause of pathologies in the long term.

The potential market for these systems is enormous. The number of deliveries in Italy is nearly 500.000 every year and the public healthcare system foresees a minimum of 3 monitoring sessions for each pregnancy. However, a single exam in the last quarter is often not enough to completely assess the situation and take appropriate decisions. So the number of monitoring sessions in Italy can be higher than 1.5 M per year.

The most common exam for evaluating the fetal wellbeing is Cardiotocographic (CTG) monitoring, which consists of measuring Fetal Heart Rate (FHR) and mother uterine contractions. In fact, fetal wellness is very strongly related with heart functioning, which makes measuring the FHR so important. CTG technology is based on a doppler ultrasound probe placed on the maternal abdomen and can only be accomplished in hospital-like structures, since it requires the supervision of expert personnel. Recognition of the echoes generated by the opening and closing of the cardiac valves of the fetus allows for the measuring of the position of the beats in time and therefore the computing of the FHR.

In the last twenty years, advances in CTG analysis have improved the quality of FHR feature extraction in order to obtain reliable indications of disease development. Great contributions arrived from advanced processing methods that provided novel parameters that were able to discriminate fetal states and healthy vs. disease groups [15,16]. These encouraging results pushed even more the research efforts to explore novel solutions towards FHR monitoring. Even if CTG is the most used technology, it still remains an indirect way to measure the fetal heartbeat. More importantly, it does not allow continuous FHR recordings, since data can only be collected at some certain fixed times, possibly separated by tenths of days if not by months.

For this reason, we started the Telefetalcare project [17], which aims to design and realize a wearable fetal tele-monitoring system that is able to collect FHR throughout the last period of the pregnancy. Its key point is the integration of a specific technology (i.e., wearable sensors and electronics) with consumer ICT devices (i.e., smartphones and tablets) and remote, high-level signal processing, data mining, and clinical decision support systems. Available methods of machine learning and artificial intelligence, applied to FHR, can be employed to improve diagnosis and prediction in pregnancy healthcare processes [18]. Furthermore, the large amount of available data already classified in CTG analysis, both in normal and pathological fetuses, ensures a robust (solid) clinical reference for the development of a new monitoring system [19]. In fact, rich clinical classified and annotated databases can be used as a knowledge source to generate reliable decision support systems. These approaches can open a new perspective on the continuous monitoring of fetal development: further information can be extracted by introducing novel analysis tools, which are more sensitive to fetal states both in healthy and stress conditions, by increasing the length, frequency, and quality of monitoring session.

Telefetalcare provides the opportunity to enlarge the time window of fetal and maternal data collection. It can enhance the performances of regular examinations, make treatments really personalized, and reduce the effort needed when a constant supervision is required.

2. Materials and Methods

2.1. Overview of the System

The system we are presenting here is the evolution of Telefetalcare [17], a project started in 2010 and aimed at developing a new pregnancy-wearable monitoring system, suited for domiciliary use. Although many different techniques are currently used to monitor FHR in the clinical practice (CTG, abdominal ECG, fetal scalp ECG, MECG, etc.), the primary need of realizing a wearable non-invasive and low-cost device encouraged the choice of measuring the abdominal Electrocardiogram (ECG).

Two major difficulties arise when using this technique: first, the separation of the maternal and the Fetal ECG (FECG) is compulsory, because they are both revealed through electrodes; second, the presence of the "vernix caseosa" during the 29th–31st weeks of the pregnancy makes it almost impossible to measure the FECG. The first problem has been solved through the use of multiple ECG leads and intelligent signal processing, as reported in the following, while the second still remains and limits the use of the system to the last weeks of pregnancy (32–42).

Moreover, when it comes to acquiring ECG signals, one of the problems faced by patients, as they start monitoring themselves, originates from their inability to properly position the electrodes. This happens because standard Ag/AgCl electrodes need to be correctly positioned and attached one-by-one using gel and glue, which is a challenging task for inexperienced users, such as pregnant women at home. For these reasons, we decided to design a wearable bodysuit provided with ECG textile electrodes specifically placed to measure abdominal FECG.

The Telefetalcare system encompasses a wearable unit, a compact electronic box for data preprocessing/transmission, and a smartphone/tablet that sends signals over the network to a remote diagnostic center and receives their results. The functional diagram of the whole system is shown in Figure 1.

Figure 1. The Functional Diagram of the Telefetalcare System.

2.2. The Wearable Unit

The wearable outfit consists of an elastic bodysuit made of cotton and *Lycra*^{TM} that is provided with electrodes made of conductive textile fibers that are directly intertwined in the garment. The adoption of a wearable garment makes the use of the system a lot simpler, because the pregnant woman just has to wear it without bothering about its fitting or properly attaching and deploying the electrodes. Thus, an elastic garment was fabricated that was able to fit different body sizes while preserving the relative positions of the electrodes.

The sensing contacts are made of silver yarns directly sewn within the bodysuit whose elastic properties ensure suitable contact with the patient skin. Since textile electrodes rely on polarized and capacitive coupling with the skin, the need to use conductive gel is also avoided.

The combined mother and fetus ECG signals are obtained by measuring the differential voltage between each of the 8 abdominal electrodes and the reference one. Of course, a critical decision affecting Telefetalcare entailed choosing the right number and the position of the electrodes on the garment. The driving goal has been to select a pattern that is able to properly sample the mother abdomen independently of the position of the fetus in the uterus. A decision to limit the lead number to 8 was mainly imposed for technical reasons. First of all, each lead requires a separate circuit for analog processing, and the adoption of a larger number of leads (i.e., 16 or 32) would have significantly increased the dimension and the power consumption of the portable electronics. Moreover, increasing the lead number would also have dramatically increased the amount of data to be processed and transmitted, possibly causing data loss problems over the wireless link used by the portable electronic to communicate with the smartphone. Thus, even though a larger number of leads would have improved the performance of the device, resulting in enhanced reliability in FECG extraction, 8 leads plus a reference one were deemed to be a good compromise (9 textile electrodes).

The placement of the sensors on the bodysuit has been carefully investigated in the previous prototypes of the system, because the main goal of their placement was to maximize the FECG signal with respect to the maternal one. Several configurations were tested in order to identify the best solution. In all of them, the position of the electrodes was limited to the garment surface, privileging those spots that were not particularly subject to movement artifacts.

The configuration shown in Figure 2a was used during the first experiments. The reference electrode is positioned on the right side of the maternal chest, while the remaining ones are placed just below the abdomen. This configuration was initially selected for the high level of the ECG signal (i.e., >300 mV pp). However, the presence of an electrode (i.e., the reference one) very close to the maternal heart caused the appearance of a strong maternal component in the recordings, which was very difficult to remove in the subsequent FECG extraction phase. Thus, the configuration shown in Figure 2b) was then chosen where the reference electrode is placed on the navel and the remaining 8 ones are placed around it with radial symmetry. The overall amplitude of the signal is lower, but the SNR between Fetal and Mother ECGs is good enough to allow more efficient FECG extraction and Fetal QRS peak detection.

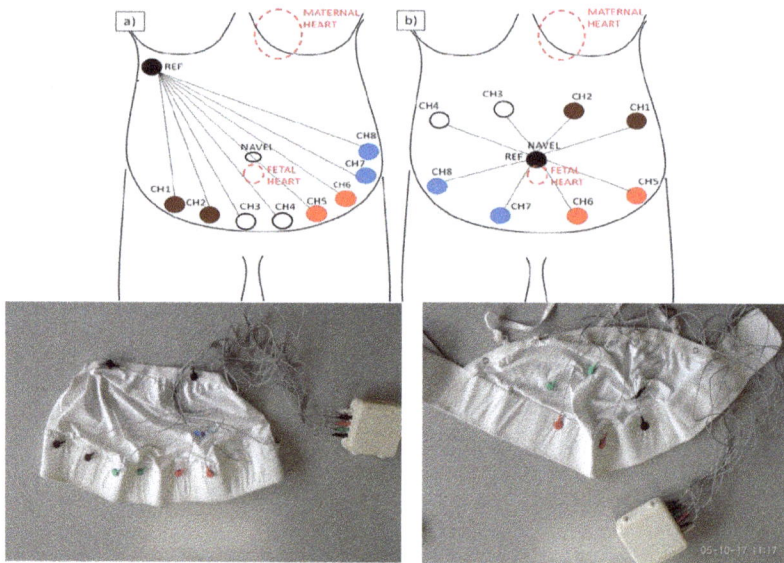

Figure 2. Relative positioning of the electrodes on the mother abdomen (**above**) and their configuration on the elastic bodysuit (**below**).

Figure 3 shows the last version of the bodysuit from the textile electrode's side, which is in contact with the maternal abdomen. These electrodes do not need the use of conductive gel and can adhere to the abdomen skin thanks to the elasticity of the bodysuit.

Figure 3. The new version of the bodysuit, viewed from the internal side.

2.3. The Electronic Box

ECG signal conditioning, A/D conversion, and wireless transmission to the smartphone using a Bluetooth (BT) link are all carried out by means of an electronic box connected to the bodysuit through standard ECG cables, as shown in Figure 4. The box was specifically designed and realized through a 3D printer, in order to contain the two small boards devoted, respectively, to analog and digital circuits.

The analog circuit for signal pre-processing operates a bandpass filter in the frequency range of 0.05–128 Hz with a controllable gain. Signal amplification was mandatory in order to exploit the full dynamic range of the Analog to Digital Converter (ADC) (i.e., 0–3.3 V), since the mother QRS peak has a range of 100 uV–1 mV, and the fetal QRS peak recorded on the abdomen has an amplitude of only 1–100 uV. Thus, the circuit was designed with a gain of approximately 2000, in order to avoid ADC saturation in case of baseline wander due to artifacts.

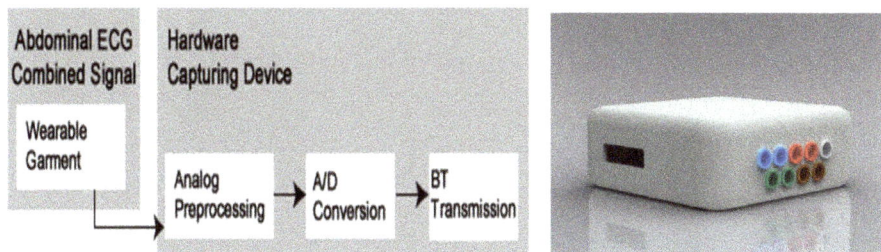

Figure 4. The block diagram of the hardware capturing device (**left**) and its box (**right**).

The FECG recorded on the abdomen has a bandwidth in the range 0.05–100 Hz, which is compatible with a sampling rate at 256 Hz imposed by constraints of the BT transmission link. In fact, the 8 channels are all sent in real time to the device, since no high-level processing takes place on the hardware device, and further increasing the sampling frequency might have caused data loss problems due to the saturation of the BT link. Data are digitally converted by a 16 bit ADC. We decided to use a 16 bit ADC because of the very low amplitude of abdominal ECG recordings. A 12 bit ADC was actually tested in an earlier prototype, but it did not provide the required signal resolution.

The first experiments were accomplished using pre-existing boards dedicated to the acquisition of electro-encephalography signals. Each of those boards supported two differential leads and was based on the chips INA118U, which includes an instrumentation amplifier, and TLC274, which has 4 operational amplifiers. Both chips are produced by Texas Instruments. The original circuit layout was not modified, but the passive components (i.e., resistors and capacitors) were changed to achieve the required bandpass range (i.e., 0.05–128 Hz) and an amplification gain of 2000.

In order to reduce the power consumption and the overall circuit dimension, a new board was designed from scratch and built based on the chips INA333 and OPA2333 also from Texas Instruments, which have much lower current profiles. Moreover, a significant difference with the preliminary board consisted in introducing a driven-leg circuit into the new one. This circuit takes the signal from the inputs, sums and negatively amplifies it, and finally reintroduces the result into the patient leg with the aim of improving the performance in terms of noise rejection and signal quality. This allows a significant reduction of the alternating current interference.

A different board was used to digitally sample the signal and send the data stream through a BT link to the external device. This board may receive up to 8 analog inputs from the analog preprocessing boards, and a 16 bit ADC with 50 µV voltage resolution samples each channel at 256 Hz. The microprocessor ARM TR711FR2 manufactured by STMicroelectronics oversees the entire functioning of the circuit. The board is powered by a 3.6 V Li-Ion rechargeable battery, similar to those used for powering mobile phones, which is also used to power up the preprocessing analog boards and guarantees up to 6 h of life.

A button controls switching on and off the entire device, and a LED is provided, which shines red when the device is powered-on and turns green during data transmission.

A further improvement of the electronics has recently been introduced by Torti et al. [20], extracting FECG on a FPGA device, which enormously decreases the amount of data transmitted via the BT connection, as explained in the next section of the paper. Unfortunately, this new approach requires a complete redesign of the electronics, and it will be included in the next release of the system.

2.4. FECG Extraction

In Abdominal FECG (AECG) recordings, the information source is FECG even though Maternal ECG (MECG) and a broad spectrum of different noises (AC interference, movement artifacts, baseline wander, electrode contact) are superimposed on it. Among disturbing noise sources, the most significant is MECG. Fetal QRS (FQRS) might be up to 10 times smaller than Maternal QRS (MQRS), thus with a very low SNR. This implies that P and T fetal ECG waves could not be reliably detected, and the only event from FECG that could be extracted after serious pre-processing is the FQRS. In addition, the classical information used in antenatal monitoring consists of FHR time series, built by identifying R-R intervals in FECG (or an equivalent as in the case of CTG).

However, in our case, FECG might not be present in some of the 8 ECG leads, depending on the position of the fetus in uterus. Thus, the FECG extraction algorithm should be able to detect when FQRS detection is successful and, in those cases, to recognize which one of the 8 ECG leads sources is the best FECG. Both noise removal and selection of the best FECG channel are extremely complex tasks to accomplish.

The extraction of Fetal ECG in the current system is performed off-line. It is organized in different steps to satisfy the following specifications: it must be (i) automatic, (ii) easy to be implemented, and (iii) working with a limited number of ECG leads (8 in our system).

Literature proposes a vast number of algorithms for Fetal and Maternal QRS identification, MECG suppression, and FECG extraction. The two main families of algorithms for FECG extraction are (i) methods based on Blind Source Separation and Independent Component Analysis [21] and (ii) methods removing maternal QRS after an averaging and subtracting procedure [22,23]. Although we explored both solutions, a method belonging to the second family is implemented in the actual release of the system. The algorithm consists of a modified and upgraded version of the Martens algorithm [24,25].

Before the algorithm application, it is necessary to submit Abdominal ECG (AECG) signals to five-step preprocessing:

(a). AECG recordings submitted to a 50 Hz digital FIR notch filter.

It removes 50 Hz power-line interference, in a very sharp and selective way. Surrounding frequency components are preserved (better than a notch analog filter).

(b). Baseline wander and high frequency noise removal from the output signal at step 1 through a FIR pass-band filter (3 to 80 Hz).

(c). Resampling of all 8 signals at 1000 Hz (T = 1 ms) in order to obtain the X_{50FILj} signals ($j = 1 \ldots 8$)

(d). X_{50FILj} signals submitted to a low-pass filter, moving-average, order 30 (like a moving window of 30 ms, with cutoff frequency 15 Hz), and results stored as X_{MAj} signals.

(e). Computation of the difference between X_{50FILj} and X_{MAj}: $Y_j = X_{50FILj} - X_{MAj}$.

Y_j signals significantly enhance the high frequency components belonging to both fetal and maternal QRSs.

2.4.1. Detection of MECG

The algorithm starts with the localization of Maternal QRSs (adapted from [24]). QRS enhancement is based on Principal Component Analysis (PCA). The input is Γ, a [8xN] data matrix containing the eight Y_j signals, each of the N samples, and the output is the first principal component C, representing the linear combination that yields maximum variance. Inter-signal correlation will be large for ECG related components, as it will be small in case of noise. Since FECG is not always present in the 8 channels, the signal C almost suppresses the FECG component.

C is divided into windows of 1 s, in order to be sure that at least a maternal QRS is contained in each window, and the absolute max (AMax) is computed in each window and temporally located. At this point, by taking a 100 ms window centered in AMax as single temporary MQRS template, the crosscorrelation between this MQRS template and the 1 s window is computed. The peak of crosscorrelation identifies the beginning of MQRS. MQRS peak is detected and located by finding the maximum within the next 100 ms after the crosscorrelation peak. Thus, by considering all windows, we obtain an array $t(k)$ containing the time locations of MQRSs.

The algorithm then goes back to the Y_j signals to extract the running MQRS template in each channel ($Temp_{MQRSj}$).

2.4.2. Construction of MQRS Template

For each Y_j signal, a window of 100 ms centered in each $t(k)$ is set ($W_{MQRSj}(k)$). A running template $Temp_{MQRSj}(k)$ is then obtained by averaging ten $W_{MQRSj}(k)$ windows preceding the actual k in the Y_j signal. The current MQRS is removed by subtracting, from the Y_j signal, the $Temp_{MQRSj}(k)$ template, scaled by a value $a_j(k)$, which minimizes mean square difference between the current QRS and the template.

$$a_j(k) = a, \text{ in which } a \rightarrow min\|W_{MQRSj}(k) - a \cdot Temp_{MQRSj}(k)\|^2$$

After averaging and MQRS template subtraction, we obtain 8 possible FECG signals:

$$F_j(i) = Y_j(i) - [a_j(k) \cdot Temp_{MQRSj}(k)](i)$$

2.4.3. Extraction of FHR Series

The extraction of FQRS from the F_js follows the same procedure applied for MQRS, except the enhancement based on PCA method, because fetal ECG is often visible in a small number (2 or 3) of abdominal leads only. Thus, no information is available about the presence of FQRSs in all 8 F_js. For this reason, FQRS detection is accomplished in all channels, despite what has been described for MQRS.

Detection of FQRS complexes on the F_js is obtained by dividing each F_j in consecutive windows of 0.5 s, finding the absolute peak in each window and considering 60 ms around it as a temporary FQRS template. The crosscorrelation between this FQRS template and the 0.5 s window is computed. As for MQRS, the crosscorrelation peak identifies the beginning of FQRS. FQRS peak is detected and located by finding the maximum within the next 60 ms after the crosscorrelation peak.

This procedure gives 8 arrays $tf_j(k)$ containing the time locations of detected FQRSs (equivalent to R-R series). Fetal R-R interval extraction is unreliable, or fetal ECG is not detectable, when inter beat distance shows values that are far from expected and/or noisy.

In order to determine if one or more $tf_j(k)$s is reliable, we compute a Quality parameter, as explained in the next paragraph. Figure 5 shows a chunk of a Y_j signal ($j = 3$) with the detection of both MQRS and FQRS.

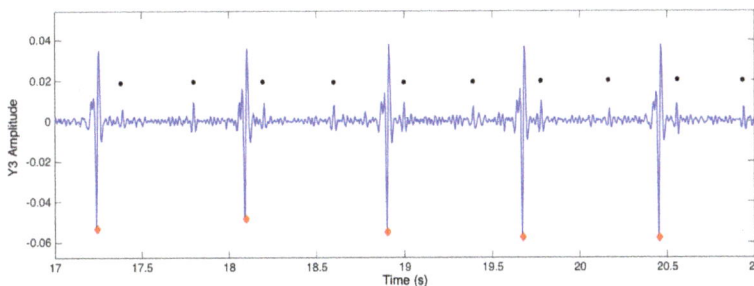

Figure 5. Maternal and Fetal ECG after the pre-processing phase ($Y_j = X_{50FILj} - X_{MAj}$, $j = 3$), in which MQRSs (red diamonds) and FQRS (black dots) are recognized and stored as $t(k)$ and $tf_j(k)$, respectively.

2.4.4. Quality Parameter

The quality parameter [17] ranks fetal QRS detection according to the reliability level of the inter beat distance.

$$RR_j(i) = tf_j(i+1) - tf_j(i)$$

The 8 Fetal RR series are ordered on the basis of the q_j index from the lowest to the highest. Low q_j values indicate good performances and allow selecting the lead (over the available 8's) with the best detection performance and the best fetal inter beat distance.

Figure 6 shows two examples of quality index.

$$q_j = mean\left(\sum_{i=1}^{M-1} \|RR_j(i+1) - RR_j(i)\|\right)$$

The computation of the q_j parameter, ranking the quality of extracted FECG and FHR, ends the algorithm application.

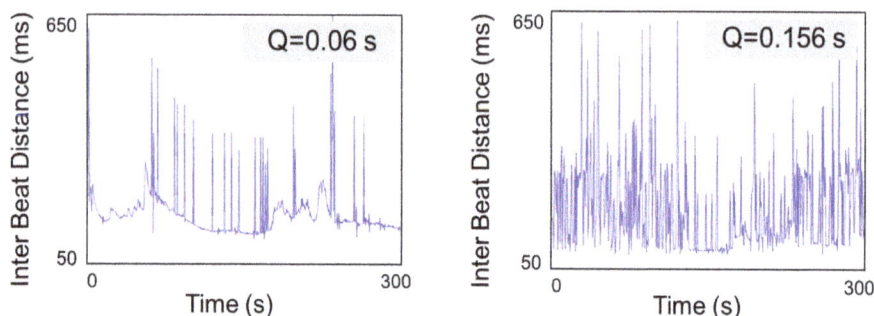

Figure 6. Two examples of quality index computation in FECG signals. Left panel show a reliable lead, and right panel refers to a lead in which the FQRS and F interbeat estimation were unreliable.

2.4.5. FECG Extraction through Field Programmable Gate Array (FPGA)

As reported in the previous subsection, an alternative technological approach to FECG extraction has been already designed and tested in our laboratory. The solution implemented by Torti et al. [20] explores the development of a special purpose custom computational unit performing fetal signal extraction to achieve lower power consumption and faster processing times in view of a potential final Application Specific Integrated Circuit (ASIC) implementation.

In this case the FPGA-based FECG extraction unit exploits a well-known blind source separation technique called Infomax [26]. It is one of the best methods in terms of reliability and documentation for implementing FECG extraction on a dedicated hardware circuit. The work in [20] describes the corresponding algorithm implementation; it produces eight different signals, which should be classified in order to correctly point out the channels related to mother (maternal signal), to fetus signal, or to noise. In addition, the classification is automatically performed and relies on a series of filter banks capable of highlighting the QRS complex, followed by different K-means instances. The architecture has been designed and implemented on an Altera Stratix V device. It is a high-end FPGA that provides the necessary hardware resources featuring relatively low power consumption. This architecture has been tested on a database made up of 343 real tracks. The results reported in [20] show that the developed architecture is real time compliant with a power consumption of about 0.5 W. In terms of elaboration speed, the proposed implementation outperforms the other solutions in the literature based on PIC, DSP and FPGA. In particular, the proposed architecture elaborates 4 s of recording in less than 30 ms. Moreover, the FPGA resource consumption of this architecture is low;

therefore, it is possible to implement other functionalities, such as data encryption for the separated tracks. On the other hand, it is also possible to lower the power consumption by adopting a less performant device, such as an Altera Cyclone V device.

This new solution implies a re-design of the electronics of the whole system, and it will be included in the next release of our system.

2.5. The Remote Monitoring Service

A fundamental element for an architecture delivering remote monitoring services is represented by the mobile device, which acts as a networking endpoint for the hardware device capturing signals. In fact, the combination of those two devices implements a remote station that may be deployed anywhere, including the patient's domicile. That station is able to capture physiological signals and send them in real time to the hospital center where they may be perused by the staff. For our project, we decided to use a commercial smartphone as the mobile device due to its low cost compared to its high-end computational capability. Moreover, the patient is expected to already own a smartphone and be comfortable with it, further reducing the economic impact of the intervention. Finally, through the use of the smartphone the patient may acquire and send data virtually anytime and anywhere, with no additional requirement other than the availability of network coverage by the phone carrier to which she is already subscribed. The smartphone operates according to a *store and forward* paradigm, so that, whenever any problem is experienced with the carrier, unsent data are never lost. Instead, they will be cached in the internal smartphone memory waiting to be transparently transmitted as soon as the network connectivity is recovered. To implement the mobile app, we chose the Android^{TM} platform because of its widespread availability on a large number of different devices with various capabilities. Furthermore, the open source approach adopted by Android^{TM} makes available a large knowledge body about its internals that is shared among enthusiast developers.

In Figure 7, we portray the architecture of the mobile app, which is modeled after a multi agent paradigm [27] and was devised based on our experience in developing similar projects [28,29]. The modularity of that framework supports the rapid prototyping of telemedicine services through the reuse of its components, thereby simplifying the switch to different medical contexts.

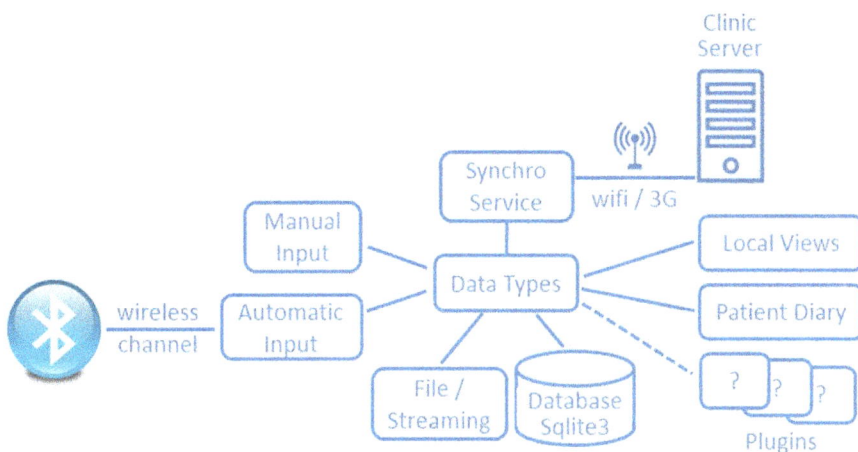

Figure 7. The computational architecture of the remote monitoring service used by Telefetalcare.

As it transpires from the figure, the whole architecture is centered on *Data Types*, which is the core component supporting the store and forward paradigm. Data are saved either exploiting the Sqlite3 database or as plain text files. In our case, data representing patient information, such as demographic data, as well as any note written by the patient herself and concerning the specific details of the acquisition to be shared with the clinic staff, are stored exploiting the Sqlite3 relational structure. Sampled values are stored instead, exploiting the file paradigm that is more efficient for long time series. Data Types interacts then with the *Synchronization* module that implements the two-way data link with the clinic server.

The *Synchronization* module exchanges data with the clinic server implementing the retry and error recovery policy. This component has been successfully exploited and validated in different areas of remote monitoring involving home patients [30,31]. Two additional modules in the architecture are represented by the *Manual Input* and *Automatic Input*. The former allows the use of the touch-screen for manually entering data, while the latter exploits BT for connecting with an external device. Finally, a set of *Plugins* are used to customize the behaviour according to a specific domain, prompting the user for input or providing specialized views over data for the chosen application (i.e., *Patient Diary* or *Local Views*). In our case, those implement the remote control for the hardware capturing device and display data to the user as shown in Figure 8, which includes a plain ECG signal.

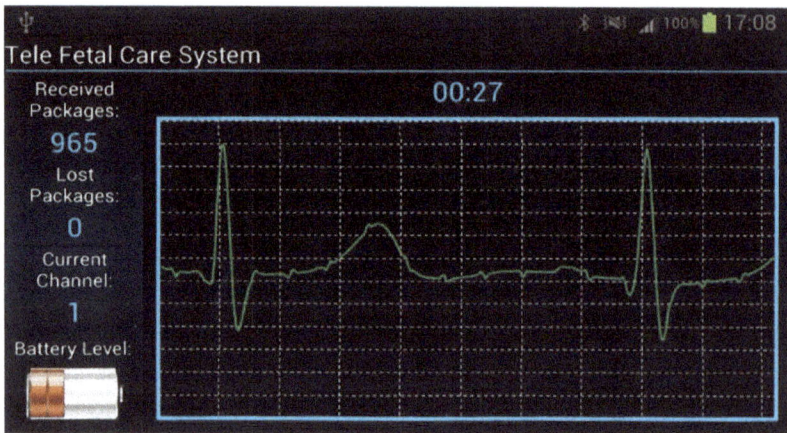

Figure 8. A snaphot of the mobile device while it is capturing a plain ECG.

At the clinic site, a server is located acting as a backend and including a synchronization engine that collects all the signal traces sent by the various patients through their apps. The server also implements the logic for extracting the FHR from the combined signals acquired and an interface for their perusal. Its operation is modeled as follows. It stores into the patient's Personal Health Record (PHR) the incoming 8 tracks that are sent after a successful recording stamped with the date-time of their acquisition. Any recording then undergoes a processing pipeline composed of three stages. The first stage includes a filter suppressing the noise superimposed to the recording, as well as the mother ECG signal, in order to emphasize the fetal component. The second pipeline stage selects the most interesting track out of the 8 available ones on which FHR will be measured, while the third one eventually accomplishes the actual FHR measurement.

The server functionality is exposed through a web application that is accessed by the doctors overseeing the service. That app includes the usual facilities for selecting patients and navigating across the sessions recorded for each one of them. Once a session is selected, the application displays the 8 tracks, as well as the computed FHR for visual inspection. All tracks may also be exported for further analysis using specialized software tools.

3. Results

The new prototype was tested on five voluntary pregnant women at the 37th week of gestation during everyday life conditions. They all gave us the informed consent for this testing phase.

3.1. Garment

All subjects reported easiness of wearing the elastic bodysuit and great comfort during the recording sessions. The abdominal ECG was recorded while subjects were sitting on a chair and reading a book. We collected more than 10 h of abdominal ECG in different sessions. The average duration of each tracing was 30 min ± 4 min.

All women had a normal course of pregnancy, and we verified "a posteriori" that they had a physiological delivery and the newborns were healthy.

In order to prepare for signal acquisition, it was only required that the subject connected the leads coming out of the sensorized garment to the electronic box, which is dedicated to analog preprocessing, analog-to-digital conversion, and digital processing. She should also turn on the smartphone, activate the Telefetalcare app installed on it, and check that the two components were successfully paired and data transmission over the BT wireless link started.

3.2. Electronics

An essential prerequisite of our solution was the reduction of any cost related to its deployment and management. To achieve this goal, we decided to devoid the electronic box acting as the signal-capturing device of any display, on the basis that smartphones already have screens smart enough to serve for that purpose. Furthermore, mobile devices make also available the primitives for easily programming their windowing system to achieve an effective user interaction through simple touch gestures. Thus, the mobile device besides offering the possibility of remotely sending the acquired signals is also used as the control unit of the hardware device. The use of BT technology allows pairing the hardware device with a wide range of receivers. Besides mobile devices, also laptops, hubs, or docking stations could be able to pair with it inasmuch they are endowed with a suitable application.

Once the mobile terminal was successfully connected through the BT link to the hardware device, the patient started the acquisition and observed in real time on its screen the signals being acquired. This feature is particularly useful, since it gives immediate feedback to the patient about the proper operation of the system. Thus, if a patient experiences any problem with one of the channels, possibly due to a lead that is not properly connected and has become noisy, the acquisition may be stopped, the lead may be repositioned, and the acquisition may be started again.

3.3. Transmission

We opted for a deferred transmission, in which the mobile terminal acts according to a store-and-forward paradigm saving a track of variable length growing at a rate of nearly 256 kBytes for every minute of sampling. When, according to the directions given by the health care staff, the expected length was achieved (usually 30 min), the patient could stop the acquisition and select whether to send that acquisition to the center, keep it locally, or discard it altogether.

3.4. Signal Quality

The quality of the electrocardiographic recordings was then compared to other commercial devices, showing good results and flat band response in the frequency range 0.1–100 Hz, as reported in [11]. During the tests, data coming from the abdominal ECG leads of the wearable garment were preprocessed and transmitted to a laptop through BT connection. A graphical user interface was developed to display signals in real time; also, on the laptop that was used to test the algorithms for the extraction of FECG from abdominal recordings, the fetal QRS and the signal classification according to the quality index were identified.

3.5. Performance of Classifiers

The performance of the actual FECG extraction algorithm was evaluated on real data acquired using Telefetalcare, achieving reliable results.

For each recording, accuracy (AC) and sensitivity (SE) of fetal and maternal QRS detection were quantified. An expert clinician evaluated the 1st-ranked traces with q value < 0.1s (according to the defined quality parameter) by counting the number of (i) QRS correctly detected (TP), (ii) missed QRS (FN), and (iii) QRS wrongly detected (FP).

AC and SE were computed in the following way:

$$AC = \frac{TP}{TP + FN + FP} \quad SE = \frac{TP}{TP + FN}$$

Classification accuracy = correct predictions/total predictions × 100.

For maternal QRS detection, we obtained an overall accuracy AC = 98.52% and sensitivity SE = 99.5%. For fetal QRS detection, overall AC = 91.26% and SE = 92.94%.

The FHR time series were then analyzed at the medical center (University of Naples Federico II) using the software we developed (2CTG2) and the most recent algorithms for the multiparametric analysis [32] and classification [18]. All subjects showed parameters belonging to the normal ranges, as we selected healthy pregnant women.

In this testing phase, we did not ask pregnant women who were at risk or had known fetal pathology to participate. For this reason, we do not report results on the classification of fetal conditions. The large amount of data obtained from the application of classification parameters to Fetal Heart Rate data from CTG recordings confirmed their ability in the fetal state classification [33]. These results were used to design the processing architecture.

The next step of our project will be to use the system in collaboration with the Obstetrics Clinic and to record tracings from a large population of pregnant women, comprising subjects at risk or with a known fetal pathology, in order to check the classifying algorithms and to generate warnings and alarms through our mobile application.

4. Discussion

In the past decade, there have been many attempts to exploit wearable monitoring systems to provide better care for patients. As a result, no convincing evidence emerged that these systems provide a cost effective solution to the problem of promptly detecting any situation that requires medical action [34]. As suggested in [35], the main reason for that is not due to intrinsic technical failures or limitations but to the underestimation of the organizational issues involved for their adoption. This becomes a sensible topic, since the majority of remote monitoring efforts have been addressing the treatment of patients affected by chronic diseases with the aim of promptly detecting any symptom calling for an action in order to avoid or decrease the onset of complications.

While there is a great potential for remote monitoring in this area [36], the long term perspective of chronic diseases, mainly involving elderly patients, poses a higher burden on the organization side. With respect to this, there is a lack of guidelines for the implementation of long term home tele-health solutions, as well as no consensus on specific clinical indicators universally trusted by clinicians drawing on data that may be automatically acquired or directly provided by patients. Those aspects, combined with the failure to account for a broader evaluation context also encompassing legal, ethical, organizational, and practical aspects, are delaying the mainstream adoption of home monitoring system for chronic patients [37].

In this context pregnancy, although it does not represent a chronic disease, can be considered like a time-bounded chronic risky condition, and home-monitoring pregnant women involves the same organizational problems as monitoring chronic patients. However, if remote monitoring is conceived of as a whole "service" comprising wearable devices for collecting signals, electronics for transmitting data, and suitable software and algorithms for analyzing data supporting the clinician specialist to take decisions on the basis of the analyzed data, the likelihood of achieving a successful "tele-monitoring" system increases enormously.

The paper describes the design and development of an end-to-end system for monitoring FHR during pregnancy. The whole system relies on a wearable sensorized garment for acquiring abdominal FECG recordings, a telemedicine module to send data to a clinic center, and a clinic server equipped with software applications able to perform an advanced quantitative analysis of the FHR variability signal through non-linear and soft computing algorithms to obtain the best classification of the fetal condition. The remote wearable device can be used by pregnant women at home without the need for external support by clinicians.

On the technical side, a recent work by Jezewski et al. [38] demonstrated the equivalence of abdominal FECG and Doppler Ultrasound (CTG) methods, in terms of ability of clinical parameters to distinguish between normal and pathological fetuses. This fact supports the remote use of advanced signal processing methods already developed for CTG analysis antepartum in the clinical center deputed to the tele-monitoring service [13]. Moreover, a recent review paper on telemonitoring in obstetrics [39] shows "the added value, for both mother and child, of telemonitoring used in prenatal follow-up program" and suggests that it can be recommended for pregnant women at risk.

Our system, within a limited set of pregnant women, demonstrated both accuracy in measuring and extracting FHR and comfort for the patients, who can check the fetal condition while staying at home by means of a low cost wearable device and a reliable telemedicine service.

Author Contributions: All authors equally contributed to this paper. In particular: Conceptualization, M.G. Signorini and G. Magenes; Methodology, M.G. Signorini, G. Lanzola, A. Fanelli, E. Torti and G. Magenes; Hardware, A. Fanelli, G. Magenes; Software, G. Lanzola, A. Fanelli and E. Torti.; Validation, M.G. Signorini, G. Magenes, A. Fanelli.; Formal Analysis, A. Fanelli, M.G. Signorini; Investigation, G. Magenes; Resources, M.G. Signorini and G. Magenes.; Data Curation, A. Fanelli.; Writing-Original Draft Preparation, G. Magenes, M.G. Signorini, G. Lanzola, E. Torti; Writing-Review & Editing, G. Magenes, G. Lanzola, M.G. Signorini; Visualization, G. Magenes, G. Lanzola.; Supervision, G. Magenes.; Project Administration, M.G. Signorini.; Funding Acquisition, M.G. Signorini and G. Magenes.

Funding: This research was funded by Italian PRIN National Project prot. N. 2008TERW82 Ministry of Education and Research.

Acknowledgments: The authors are grateful to Dr. Marta Campanile, Ob-Gyn Clinics, University of Naples, for the possibility of recording patients in her laboratory.

Conflicts of Interest: The authors declare no conflict of interest.

References

1. Norris, A. *Essentials of Telemedicine and Telecare*; John Wiley & Sons: Hoboken, NJ, USA, 2001.
2. Losiouk, E.; Lanzola, G.; Del Favero, S.; Boscari, F.; Messori, M.; Rabbone, I.; Bonfanti, R.; Sabbion, A.; Iafusco, D.; Schiaffini, R.; et al. Parental evaluation of a telemonitoring service for children with Type 1 Diabetes. *J. Telemed. Telecare* **2017**. [CrossRef] [PubMed]
3. Gaikwad, R.; Warren, J. The role of home-based information and communications technology interventions in chronic disease management: A systematic literature review. *Health Inform. J.* **2009**, *15*, 122–146. [CrossRef] [PubMed]
4. Martirosyan, M.; Caliskan, K.; Theuns, D.A.M.J.; Szili-Torok, T. Remote monitoring of heart failure: Benefits for therapeutic decision making. *Expert Rev. Cardiovasc. Ther.* **2017**, *15*, 503–515. [CrossRef] [PubMed]
5. Farnia, T.; Jaulent, M.C.; Steichen, O. Evaluation Criteria of Noninvasive Telemonitoring for Patients with Heart Failure: Systematic Review. *J. Med. Internet Res.* **2018**, *20*, e16. [CrossRef] [PubMed]
6. Schoenfeld, M.H.; Compton, S.J.; Mead, R.H.; Weiss, D.N.; Sherfesee, L.; Englund, J.; Mongeon, L.R. Remote monitoring of implantable cardioverter defibrillators: A prospective analysis. *Pacing Clin. Electrophysiol.* **2004**, *27* Pt 1, 757–763. [CrossRef] [PubMed]

7. Varma, N.; Michalski, J.; Epstein, A.E.; Schweikert, R. Automatic remote monitoring of implantable cardioverter-defibrillator lead and generator performance. *Circ. Arrythmia Electrophysiol.* **2010**, *3*, 428–436. [CrossRef] [PubMed]

8. Lanzola, G.; Scarpellini, S.; Di Palma, F.; Toffanin, C.; Del Favero, S.; Magni, L.; Bellazzi, R. Monitoring Artificial Pancreas Trials through Agent-based Technologies. *J. Diabetes Sci. Technol.* **2014**, *8*, 216–224. [CrossRef] [PubMed]

9. Di Rienzo, M.; Meriggi, P.; Rizzo, F.; Castiglioni, P.; Lombardi, C.; Ferratini, M.; Parati, G. Textile technology for the vital signs monitoring in telemedicine and extreme environments. *IEEE Trans. Inf. Technol. Biomed.* **2010**, *14*, 711–717. [CrossRef] [PubMed]

10. Curone, D.; Secco, E.L.; Tognetti, A.; Loriga, G.; Dudnik, G.; Risatti, M.; Whyte, R.; Bonfiglio, A.; Magenes, G. Smart garments for emergency operators: The proetex project. *IEEE Trans. Inf. Technol. Biomed.* **2010**, *14*, 694–701. [CrossRef] [PubMed]

11. Vainoras, A.; Marozas, V.; Korsakas, S.; Gargasas, L.; Siupsinskas, L.; Miskinis, V. Cardiological telemonitoring in rehabilitation and sports medicine. *Stud. Health Technol. Inform.* **2004**, *105*, 121–130. [PubMed]

12. Giansanti, D.; Maccioni, G.; Macellari, V.; Mattei, E.; Triventi, M.; Censi, F.; Calcagnini, G.; Bartolini, P. A novel, user-friendly step counter for home telemonitoring of physical activity. *J. Telemed. Telecare* **2008**, *14*, 345–348. [CrossRef] [PubMed]

13. Di Lieto, A.; De Falco, M.; Campanile, M.; Papa, R.; Torok, M.; Scaramellino, M.; Pontillo, M.; Pollio, F.; Spanik, G.; Schiraldi, P.; et al. Four years' experience with antepartum cardiotocography, using telemedicine. *J. Telemed. Telecare* **2006**, *12*, 228–233. [CrossRef] [PubMed]

14. Buysse, H.; De Moor, G.; Van Maele, G.; Baert, E.; Thienpont, G.; Temmerman, M. Cost-effectiveness of telemonitoring for high-risk pregnant women. *Int. J. Med. Inform.* **2008**, *77*, 470–476. [CrossRef] [PubMed]

15. Tagliaferri, S.; Fanelli, A.; Esposito, G.; Esposito, F.G.; Magenes, G.; Signorini, M.G.; Campanile, M.; Martinelli, P. Evaluation of the Acceleration and Deceleration Phase-Rectified Slope to Detect and Improve IUGR Clinical Management. *Comput. Math. Methods Med.* **2015**. [CrossRef] [PubMed]

16. Hoyer, D.; Żebrowski, J.; Cysarz, D.; Gonçalves, H.; Pytlik, A.; Amorim-Costa, C.; Bernardes, J.; Ayres-de-Campos, D.; Witte, O.W.; Schleußner, E.; et al. Monitoring fetal maturation-objectives, techniques and indices of autonomic function. *Physiol. Meas.* **2017**, *38*, R61–R88. [CrossRef] [PubMed]

17. Fanelli, A.; Signorini, M.G.; Ferrario, M.; Perego, P.; Piccini, L.; Andreoni, G.; Magenes, G. Telefetalcare: A first prototype of a wearable fetal electrocardiograph. In Proceedings of the IEEE Engineering in Medicine and Biology Society Conference, EMBC 2011, Boston, MA, USA, 30 August–3 September 2011; pp. 6899–6902.

18. Magenes, G.; Bellazzi, R.; Malovini, A.; Signorini, M.G. Comparison of data mining techniques applied to fetal heart rate parameters for the early identification of IUGR fetuses. In Proceedings of the IEEE Engineering in Medicine and Biology Society EMBC 2016, Orlando, FL, USA, 16–20 August 2016; pp. 916–919. [CrossRef]

19. Giuliano, N.; Annunziata, M.L.; Esposito, F.G.; Tagliaferri, S.; Di Lieto, A.; Magenes, G.; Signorini, M.G.; Campanile, M.; Arduini, D. Computerised analysis of antepartum foetal heart parameters: New reference ranges. *J. Obstet. Gynaecol.* **2017**, *37*, 296–304. [CrossRef] [PubMed]

20. Torti, E.; Koliopoulos, D.; Matraxia, M.; Danese, G.; Leporati, F. Custom FPGA processing for real-time fetal ECG extraction and identification. *Comput. Biol. Med.* **2017**, *80*, 30–38. [CrossRef] [PubMed]

21. Zarzoso, V.; Nandi, A.K. Noninvasive fetal electrocardiogram extraction: Blind separation versus adaptive noise cancellation. *IEEE Trans. Biomed. Eng.* **2001**, *48*, 12–18. [CrossRef] [PubMed]

22. Cerutti, S.; Baselli, G.; Civardi, S.; Ferrazzi, E.; Marconi, A.M.; Pagani, M.; Pardi, G. Variability analysis of fetal heart rate signals as obtained from abdominal electrocardiographic recordings. *J. Perinat. Med.* **1986**, *14*, 445–452. [CrossRef] [PubMed]

23. De Lathauwer, L.; De Moor, B.; Vandewalle, J. Fetal electrocardiogram extraction by blind source subspace separation. *IEEE Trans. Biomed. Eng.* **2000**, *47*, 567–572. [CrossRef] [PubMed]

24. Martens, S.M.M.; Mischi, M.; Oei, S.G.; Bergmans, J.W.M. An Improved Adaptive Power Line Interference Canceller for Electrocardiography. *IEEE Trans. Biomed. Eng.* **2006**, *53*, 2220–2231. [CrossRef] [PubMed]

25. Martens, S.M.; Rabotti, C.; Mischi, M.; Sluijter, R.J. A robust fetal ECG detection method for abdominal recordings. *Physiol. Meas.* **2007**, *28*, 373–388. [CrossRef] [PubMed]

26. Bell, J.; Sejnowski, T.J. An information maximization approach to blind separation and blind deconvolution. *Neural Comput.* **1995**, *7*, 1129–1159. [CrossRef] [PubMed]

27. Capozzi, D.; Lanzola, G. An agent-based architecture for home care monitoring and education of chronic patients. In Proceedings of the IEEE Conference Complexity in Engineering, COMPENG'10, Rome, Italy, 22–24 February 2010; pp. 138–140. [CrossRef]

28. Lanzola, G.; Capozzi, D.; D'Annunzio, G.; Ferrari, P.; Bellazzi, R.; Larizza, C. Going mobile with a multiaccess service for the management of diabetic patients. *J. Diabetes Sci. Technol.* **2007**, *1*, 730–737. [CrossRef] [PubMed]

29. Lanzola, G.; Ginardi, M.G.; Mazzanti, A.; Quaglini, S. Gquest: Modeling patient questionnaires and administering them through a mobile platform application. *Comput. Methods Programs Biomed.* **2014**, *117*, 277–291. [CrossRef] [PubMed]

30. Losiouk, E.; Lanzola, G.; Galderisi, A.; Trevisanuto, D.; Steil, G.M.; Facchinetti, A.; Cobelli, C. A telemonitoring service supporting preterm newborns care in a neonatal intensive care unit. In Proceedings of the 3rd IEEE International Forum on Research and Technologies for Society and Industry (RTSI) 2017, Modena, Italy, 11–13 September 2017. [CrossRef]

31. Capozzi, D.; Lanzola, G. A generic telemedicine infrastructure for monitoring an artificial pancreas trial. *Comput. Methods Programs Biomed.* **2013**, *110*, 343–353. [CrossRef] [PubMed]

32. Fanelli, A.; Magenes, G.; Campanile, M.; Signorini, M.G. Quantitative assessment of fetal well-being through CTG recordings: A new parameter based on phase-rectified signal average. *IEEE J. Biomed. Health Inform.* **2013**, *17*, 959–966. [CrossRef] [PubMed]

33. Signorini, M.G.; Fanelli, A.; Magenes, G. Monitoring fetal heart rate during pregnancy: Contributions from advanced signal processing and wearable technology. *Comput. Math. Methods Med.* **2014**, 707581. [CrossRef] [PubMed]

34. Whitten, P.S.; Mair, F.S.; Haycox, A.; May, C.R.; Williams, T.L.; Hellmich, S. Systematic review of cost effectiveness studies of telemedicine interventions. *Br. Med. J.* **2002**, *324*, 1434–1437. [CrossRef]

35. Hardisty, A.R.; Peirce, S.C.; Preece, A.; Bolton, C.E.; Conley, E.C.; Gray, W.A.; Rana, O.F.; Yousef, Z.; Elwyn, G. Bridging two translation gaps: A new informatics research agenda for telemonitoring of chronic disease. *Int. J. Med. Inform.* **2011**, *80*, 734–744. [CrossRef] [PubMed]

36. Pare, G.; Jaana, M.; Sicotte, C. Systematic review of home telemonitoring for chronic diseases: The evidence base. *J. Am. Med. Inform. Assoc.* **2007**, *14*, 269–277. [CrossRef] [PubMed]

37. Koch, S. Home telehealth current state and future trends. *Int. J. Med. Inform.* **2006**, *75*, 565–576. [CrossRef] [PubMed]

38. Jezewski, J.; Wrobel, J.; Matonia, A.; Horoba, K.; Martinek, R.; Kupka, T.; Jezewski, M. Is Abdominal Fetal Electrocardiography an Alternative to Doppler Ultrasound for FHR Variability Evaluation? *Front. Physiol.* **2017**, *16*, 305. [CrossRef] [PubMed]

39. Lanssens, D.; Vandenberk, T.; Thijs, I.M.; Grieten, L.; Gyselaers, W. Effectiveness of Telemonitoring in Obstetrics: Scoping Review. *J. Med. Internet Res.* **2017**, *19*, e327. [CrossRef] [PubMed]

![technologies logo] *technologies*

MDPI

Article

Wearable Inertial Sensing for ICT Management of Fall Detection, Fall Prevention, and Assessment in Elderly

Vincenzo Genovese, Andrea Mannini *, Michelangelo Guaitolini and Angelo Maria Sabatini

The Biorobotics Institute Scuola Superiore Sant'Anna, 56127 Pisa, Italy;
vincenzo.genovese@santannapisa.it (V.G.); michelangelo.guaitolini@santannapisa.it (M.G.);
sabatini@santannapisa.it (A.M.S.)
* Correspondence: andrea.mannini@santannapisa.it; Tel.: +39-050-88-3410

Received: 12 September 2018; Accepted: 1 October 2018; Published: 2 October 2018

Abstract: Falls are one of the most common causes of accidental injury: approximately, 37.3 million falls requiring medical intervention occur each year. Fall-related injuries may cause disabilities, and in some extreme cases, premature death among older adults, which has a significant impact on health and social care services. In recent years, information and communication technologies (ICT) have helped enhance the autonomy and quality of life of elderly people, and significantly reduced the costs associated with elderly care. We designed and developed an integrated fall detection and prevention ICT service for elderly people, which was based on two wearable smart sensors, called, respectively, WIMU fall detector and WIMU data-logger (Wearable Inertial Measurement Unit, WIMU); their goal was either to detect falls and promptly react in case of fall events, or to quantify fall risk instrumentally. The WIMU fall detector is intended to be worn at the waist level for use during activities of daily living; the WIMU logger is intended for the quantitative assessment of tested individuals during the execution of clinical tests. Both devices provide their service in conjunction with an Android mobile device. The ICT service was developed and tested within the European project I-DONT-FALL (Integrated prevention and Detection sOlutioNs Tailored to the population and risk factors associated with FALLs, funded by EU, action EU CIP-ICT-PSP-2011-5: GA #CIP-297225). Sensor description and preliminary testing results are provided in this paper.

Keywords: fall detection; fall prevention; ICT; inertial sensing

---

1. Introduction

Falls represent a major public health problem requiring medical attention: among older adults (over 65), one in four falls annually, and one dies every 19 min as result of falling [1]. Falling has significant consequences which affect the quality of life in the elderly, because falls can dramatically change an elderly person's self-confidence and motivation, thereby affecting their ability to live independently in a dramatically vicious cycle which tends to worsen with age. Even though human beings are all at risk of falling, some factors such as age, gender, health conditions, and the history of previous falls show remarkably high correlations to the type and severity of injuries which occur as a result of falling [2]. Among healthy people, adults over 65 years of age experience high risk of serious injury due to falling, and this risk keeps increasing with age. In general terms, the risk of fall is related to extrinsic (environment-dependent) and to intrinsic factors (physical, sensory and cognitive changes typical of ageing); moreover, fear of falling is related to adopting overly-cautious gait habits [2], and this might in turn cause falls that result in increased risk of falls, fear of falling, and functional decline [3]. As a result, fall prevention activities are carried out across a range of health disciplines including occupational therapy, physiotherapy, general practice, nursing, geriatric, gerontology health, and social care [4–6]. Fall risk assessment concerns the evaluation of risk factors.

Relevant efforts in mitigating risk factors are targeted to physiological factors, which include muscle strength and balance, stability, posture, and gait reaction time [7].

Fall risk definition and instrumental assessment are relevant for adapting the level of assistance needed by elderly people and tailoring preventive measures to specific subjects that are deemed to be at high risk of falling. The risk of falling is generally evaluated using questionnaires, despite their associated problems of subjectivity and limited accuracy [8]. A number of experimental tests (e.g., Berg balance scale, Timed Up and Go, Turn 180° test) have been developed to screen older people in the community or in a clinical setting [7]; risk can also be evaluated by clinical and functional assessment including posture and gait, independence in daily life, cognition, and vision [8,9]. However, previous studies still report limitations in accuracy and versatility, preventing routinely use of this process in clinical practice [8].

Recently, information and communication technologies (ICT) have been increasingly applied in the attempt to improve the level of autonomy and quality of life of elderly people at risk of falling, and a plethora of services were considered for fall prevention and detection, in the latter case, with the possibility to send alarms and call for rescue once falls have been detected [8,9]. Technology-based interventions have been deployed in a wide range of contexts, and include: (i) diagnosing and treating fall risks [10,11]; (ii) increasing adherence to interventions [12,13]; (iii) detecting falls and alerting caregivers or next of kin [14,15]. ICTs could also play a key role in enabling older adults to self-assess their fall risk, reducing costs and lessening the burden on the healthcare system, whilst also improving the quality and effectiveness of care provided [16]. Despite the abundance of ICT systems, and the availability of several interesting implementations, e.g., [17–21], there are still several challenges that may potentially impact their use in practice [22].

This paper describes the design, development, and preliminary testing of a custom wearable sensing device and ICT service for fall detection and fall risk assessment as an element of the I-DONT-FALL platform for fall risk assessment and mitigation. The European project I-DONT-FALL (Integrated prevention and Detection sOlutioNs Tailored to the population and risk factors associated with FALLs, CIP-297225) aims at pursuing a multi-factorial approach to fall detection and fall risk assessment and mitigation.

The developed ICT service was based on two wearable smart sensors, called, respectively, WIMU fall detector and WIMU data-logger (Wearable Inertial Measurement Unit, WIMU); their goal was either to detect falls and promptly react in case of fall events, or to quantify fall risk instrumentally, in line with other proposed solutions [15–18]. The WIMU fall detector was intended to be worn at the waist level for use during activities of daily living; the WIMU logger was intended for the quantitative assessment of tested individuals during the execution of clinical tests. Both devices provide their service in conjunction with an Android mobile device.

An interesting feature of the I-DONT-FALL approach is that the developed ICT services are intended to be used not only to automatically detect falls, as most competing solutions do, but also to prevent falls. Fall prevention is intended in this context to provide support to clinical decisions regarding the fall risk profile of patients, whose gait is assessed using wearable sensor technologies. In regard to the problem of fall detection, we have previously developed a method for pre-impact fall detection based on sensor fusion algorithms combining data from a waist-based inertial measurement unit integrating an air pressure sensor, [23]. There, a thorough discussion of the advantages of using air pressure sensors in a fall detector was done, and our results were compared to those achieved by alternative implementations of state-of-the-art waist-based fall detectors. As for the preliminary testing of our approach concerning fall prevention, we describe in this paper our solution to the problem of getting an instrumental version of a widely used clinical test, namely the Six Minutes Walking Test (6MWT).

The 6MWT was administered to both high and low fall-risk patients wearing the proposed system at the waist level. Waist movement during walking plays a critical role in successful locomotion, and contributes to gait stability among older people [24]. In the preliminary tests, the acquired signals were processed to extract gait parameters: stride time, cadence, stride time variability, root mean square of the vertical acceleration, and harmonic ratios (HR) of the three components of the acceleration. In fact, these gait variables, or a subset of them, have been previously considered in several research reports [25–27], sometimes with a specific focus on the study of falls [26]. Another parameter of interest in 6MWT was the walked distance which was not computed using the WIMU sensor data, but was measured manually. Walked distance can be considered a proxy to walking speed, which can be estimated as the ratio between the distance and the time elapsed during the test. Recently, solutions have been proposed concerning the application of smartphone technology to the development of a practical and easy-to-use tool for rehabilitation professionals to use in the management of 6MWT [28]. While a similar approach could have been considered here, we preferred to opt for the use of an external sensor unit, namely the WIMU. This was pursued with the aim of extending the number and type of gait variables that could be involved in the assessment (in particular, the HRs), and to be compliant with the experimental setups that are described in the literature [25,26,28].

2. Materials and Methods

2.1. The Overall ICT Solution

The I-DONT-FALL project aims at implementing a fall detection and risk management framework for a multi-factorial approach to risk assessment [29]. The fall detection service is depicted in Figure 1. The patient wears a fall detector, namely a smart sensor that is capable of identifying fall-related impacts. When a fall event is detected, an alarm is issued via the Bluetooth (BT) connection between the fall detector and an Android device from which it is sent to a call center and to a repository of phone numbers and/or e-mail addresses (i.e., relatives and caregivers). The Android device automatically selects the channel (Wi-Fi and/or GPRS/3G-4G) over which the alarm is to be broadcast. The call center closes the loop by contacting the patient/caregiver to check on his/her conditions and to acquire context information about the fall event. Context information is added to the sensor data in order to provide a clear representation of the fall event even in the case of false positives. Sensing data corresponding to 2.5-s long windows that are centered around the fall time of occurrence are automatically stored on the smartphone and transferred to a secure server. The activation of the call center is based on machine-machine interaction provided by a web service. All the gathered data from different users in different locations are stored on a remote secure server. The web service structure information is provided by the SOAP (Simple Object Access Protocol) protocol.

Three general methods for read access to patient data on the secure server by a third-party healthcare system are provided:

- access to HTML/Java streams that allow interactive, top-down browsing of patient data, using a unique, time-limited URL;
- access to PDF format summary reports of the patient data covering a specified period of time, normally 28-day blocks;
- subscription to specific types of low-level physiological data and interventions that have been recorded for or by the patient within the system.

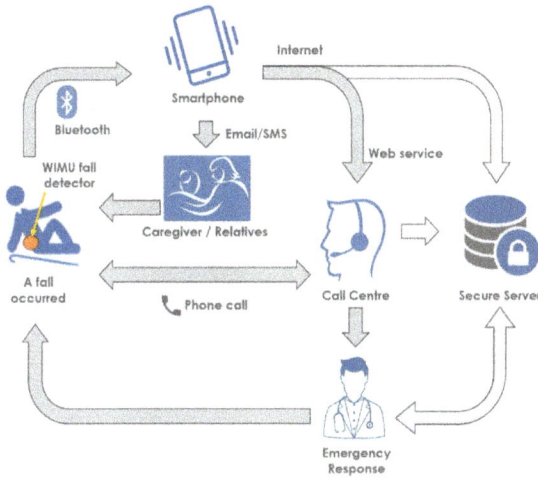

Figure 1. Fall detection service architecture: in case of an alarm by the WIMU fall detector, accelerometer data and the alarm itself are sent to the smartphone by Bluetooth connection. The smartphone automatically sends warning messages to a list of recipients (SMS/email) and alerts the call center that directly contacts the elderly individual on the smartphone, and if needed, clinical emergency staff. In a secondary loop, information about the fall event are uploaded to a secure server automatically by the smartphone and manually by the call center and clinical staff that are also allowed to access the information in the patient's record.

2.2. The Hardware

The fall detector is a battery-powered (3.7 V Li-polymer Rechargeable) device (Figure 2) intended to be worn around the waist (weight 100 g) during indoor daily activities; the battery life is about 20 h, and data from the internal sensors is sampled at a rate of 100 Hz.

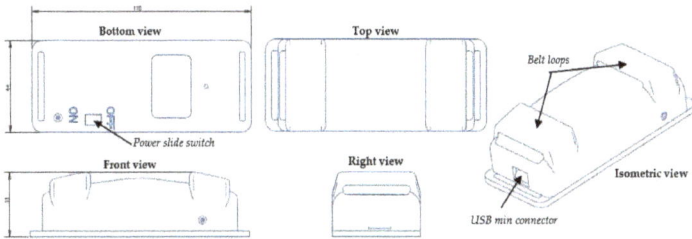

Figure 2. The WIMU fall detector case with its dimensions in different views (units in mm).

The electronics of the fall detector embeds two stacked Printed Circuit Boards (PCBs) of identical size: the controller and the sensor board. The controller is a commercial device based on a 32-bit low-power ARM cortex M0 Core running at 48 MHz, 8 KB RAM, 32 KB FLASH: the NXP LPC11U24. The software running on the controller is developed using an online Software Development Kit (SDK) that supports a C/C++ programming environment including libraries for peripheral abstraction.

The sensor board is a two-layer PCB (Figure 3), specifically designed and developed to host a battery charger powered connecting the device to an USB port via its mini USB connector, a micro SD card slot for internal logging, a Bluetooth module (RN-42 by Roving Networks), and four sensors. In particular, one tri-axial accelerometer (BMA180 by Bosch Sensortec), one tri-axial gyroscope (ITG-3200 by InvenSense), one tri-axial magnetic sensor (HMC5883L by Honeywell), and a barometric pressure sensor (BMP085 by Bosch), all sample at a rate of 100 Hz. The sensor setup is shown in Table 1.

Figure 3. The sensor board (layer 1 and 2).

Table 1. WIMU sensor board setup.

Device	Low Pass Filter Cut Off Frequency (Hz)	Sensing Range	Sensitivity-Resolution
Accelerometer	1200	$\pm 4\,g$ ($1\,g = 9.81$ m/s^2)	0.000488 g/LSB
Gyroscope	256	$\pm 2000\,°/s$	0.0696 ($°$/s)/LSB
Magnetometer		± 0.88 Gauss	0.00729 Gauss/LSB
Pressure sensor		+9000 m, ..., 500 m above the sea level	0.01 hPa

The logger is a battery-powered (3.7V Li-polymer Rechargeable) device (Figure 4) intended to be worn around the waist (weight: 90 g) during the execution of specific tests by the patients. The battery life is about 4 h.

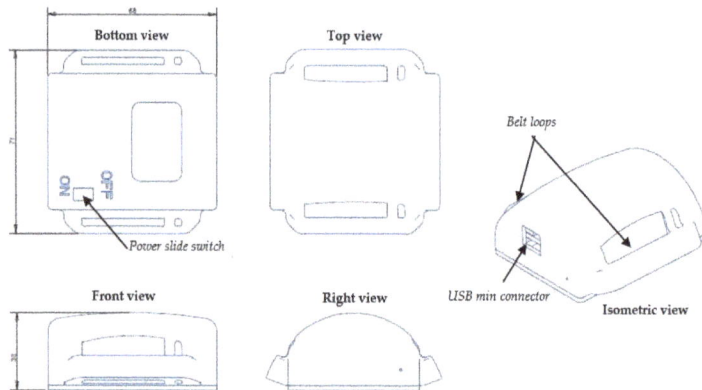

Figure 4. The WIMU logger case with its dimensions in different views (units in mm).

The logger electronics differ from the those of the fall detector only in the features of the controller. The controller is now a commercial device based on a 32-bit ARM cortex M3 core running at 96 MHz, 32 KB RAM, 512 KB FLASH: the NXP LPC1768. Basically, the difference between the NXP LPC1768 and the NXP LPC11U24 is in the power consumption and memory capabilities. Computational and memory capabilities are higher in this controller compared to those of the fall detector. The sensor board is identical to the one used for the fall detector.

2.3. The Fall Detector: Functionalities

The fall detector is designed to achieve a continuous-data monitoring for almost the full length of the day. Working in conjunction with an Android mobile device and according to the diagram of Figure 5, the fall detector provides the following functions:

- **automatic fall detection**—a built-in algorithm of fall detection identifies impacts that may be related to falls, generates and displays alarms, and sends them to the Android device using BT connectivity;
- **automatic logging** (and subsequent transfer to the Android device)—acceleration data occurring in a time window around the fall event are stored; this is a useful feature to keep track of fall events and use the logged data for post-hoc processing, e.g., further refinement and tuning of the built-in fall detection algorithm;
- **parsing and managing of a list of commands** coming from the mobile device to change the parameters of the algorithms, obtain battery information, and troubleshoot possible device errors.
- **estimation of the activity level (AL)**—the built-in algorithm of AL estimation helps keep track of the intensity of the physical activity of the patient during the day.

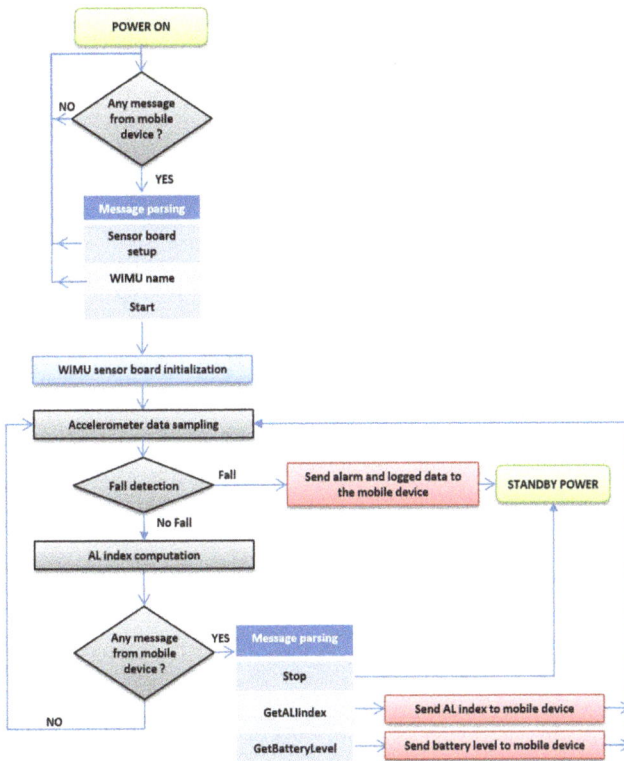

Figure 5. Fall detector flow diagram.

As highlighted in the flowchart in Figure 5, after the system start-up, the fall detector waits for commands coming from the Android device. After receiving a "Start" command, the sensing board is initialized and a control loop is activated with the purpose of sampling the accelerometer data, executing the fall detection algorithm, computing the AL index, and parsing of the commands coming from the mobile device. In case a fall event is detected, an alert message is sent to the mobile device and the WIMU is frozen in power standby. The commands available to the mobile device allow the control loop exiting for the WIMU standby (Stop command) and the collection of data about the battery level and the AL index (GetBatteryLevel and GetALIndex commands).

A detailed description of the fall detection and AL estimation algorithms are beyond the scope of this paper; they are reported in the technical deliverables of the I-DONT-FALL project [30].

2.4. The Data Logger: Functionalities

The main functionality of the logger is data acquisition and their transfer to the Android device. Namely, the logger collects raw data from the set of sensors embedded in the sensor board at a rate of 100 Hz in two different modes, see the flowchart in Figure 6:

1. **standalone mode**—the data are stored in the internal memory of the device, which is capable of storing records lasting six minutes;
2. **continuous mode**—the data are transferred sample-by-sample to the Android mobile device; the maximum duration of a data record will depend on the battery duration (about four hours) of the wearable device. Considering the sampling rate (100 Hz) and a sensory data frame length of 38 bytes, the max data length collected in four hours is about 60 MB.

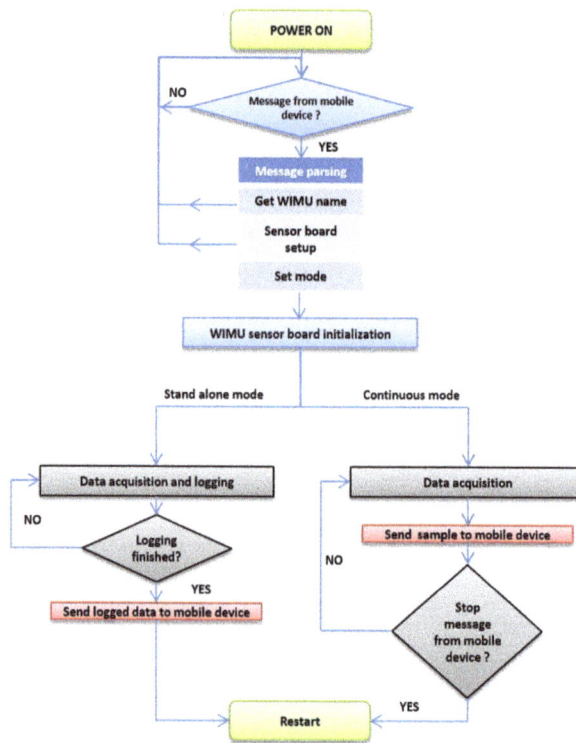

Figure 6. Logger flow diagram.

2.5. The Android Mobile Device

The mobile device runs the WIMU manager, an app which is able to operate either the fall detector or the logger. The full system operationality relies on the integrity of wireless communication (BT, GSM/GPRS/3G-4G and/or IEEE 802.11 b/g/n).

All communications between the WIMUs and the mobile device are based on the SLIP protocol (Serial Line Internet Protocol). The WIMU manager incorporates an FTP server that allows the data to be transferred to a remote data server.

When working in fall detection mode, the WIMU manager consists of four threads which work in parallel to accomplish the following tasks:

1. periodic check of the connection with WIMU;
2. periodic check of the connection with the call-center;
3. acquisition of the ADL index and WIMU's battery level;
4. alarm management;

   a. propagate the alarm to different recipients (see Table 2);
   b. upload of acceleration data logged by the WIMU in a time window around the fall event.

Table 2. Android mobile device fall detection alarm management. Tick marks indicates the used

	Internet		Telephone Network	
Alarm Recipient	**Web Service**	**e-mail**	**SMS**	**Pre-Recorded Phone Calls**
Call centre	✓			
Family and friends		✓	✓	✓

In case of connection failure or fall-detection, the Android device generates text and vocal messages. AL index and the WIMU battery level are continuously displayed on the main activity of the WIMU manager (Figure 7).

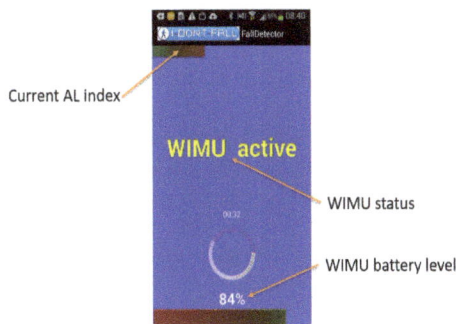

Figure 7. WIMU manager app front end.

When working in logger mode, the WIMU manager implements different functionalities:

1. the selection of standalone mode (the logging duration can be specified) or continuous mode;
2. acquisition of the WIMU battery level;
3. acquisition of the data collected by the WIMU (at the end of the experiment for the standalone mode, sample-by-sample in continuous mode);

4. calculation and displaying of the Fast Fourier Transform (FFT) of the data acquired by the WIMU during a diagnostic test data acquisition. This tool is useful to let the user periodically check the correct functionality of the instrument.

2.6. Instrumentation of the Six-Minutes Walking Test

To test the potentialities of the proposed technological solution, an instrumental version of the 6MWT has been proposed by measuring lower trunk accelerations using the WIMU, configured as a wireless data-logger. In particular, the results of a preliminary experimental session involving elderly subjects are reported to better explicate the potentialities of the proposed technological solution. Two groups were tested for this preliminary evaluation of the hardware, involving a subset of the overall IDONTFALL dataset, including 25 high fall-risk and 25 age-matched low fall-risk patients respectively, grouped and selected by the clinical partners of the I-DONT-FALL project [31]. The WIMU was mounted at the L3 spinous process (lower trunk) using a Velcro belt carefully placed in order to not restrict the subjects' movement. Subjects were asked to stand still in their upright posture for few seconds before starting the test. The 6MWT was administered along a corridor 32-m long (approximately) with two small cones on the floor to mark the start and end points of each trip. No specific instructions were given to the operators to calibrate the WIMU sensors; care was required in fixing the WIMU to the body to align the device local frame to the anatomical axes. Trunk linear accelerations were measured along the vertical (VT), anteroposterior (AP), and medio-lateral (ML) axes, sampled at 100 Hz. At the end of each test, collected data were uploaded to the Android smartphone via Bluetooth.

The first gait variable considered for the 6MWT assessment is the walking distance D, expressed in meters, and measured manually by the experimenter. Stride time, T, expressed in seconds, is the time elapsed from the first contact of the (right) leg with the ground to the next; stride time variability is computed from the coefficient of variation (COV, standard deviation divided by mean value $\times 100$) of stride time, which quantifies variability while taking into account mean performance. These gait variables, and especially stride time variability, were considered in past studies [7]. The issue of gait variability is challenging, since variability in motor function can be regarded either as a marker of impaired motor control or as a positive sign of system adaptation [32]. Gait involves cycles that are characterized by regularity, but also balance components characterized by variability. Identifying such components, while taking into account noise random error, is a challenge. If random error is large, reliability will be low, and any variable will show large variability. Thus, variability caused by random error may be misinterpreted as an indicator of adaptability or impairment. The estimation of spatio-temporal gait parameters requires the detection of subsequent foot contacts (onset and end of stride cycles). Several studies have addressed the relationship between measured accelerations (on trunk, thigh, shank and foot) and spatio-temporal gait parameters. We adapted the method proposed by Zijlstra and Hof [25] for stride time determination.

The harmonic ratios (HRs) are dimensionless quantities that, as they are derived from body accelerations, offer insights into the underlying mechanisms of balance control during gait [33]. HRs provide information on the ability of subjects to control their trunk smoothly during walking, providing an indication of whole-body balance and coordination (gait stability). We measure HRs in three directions, which allows us to exam directional responses. It can be speculated that movements in the ML direction during walking can be controlled differently than those in the AP plane, raising interesting questions concerning the ability of HRs to correlate with or predict falls. It is hypothesized that significantly lower HRs are found in unstable older adults (self-reported falls or unsteadiness) when compared to normal groups. Higher HRs indicate smoother and more stable trunk movement during gait [33]. Typical AP and VT acceleration patterns of the lower trunk during walking exhibit two major acceleration peaks per stride, one for each step; thus, frequency decomposition through Fourier analysis yields a dominance of the second harmonic and subsequent even harmonics. The even harmonics for the AP and VT indicate the in-phase components of the signal, whereas the odd harmonics comprise the out-of-phase components (minimized in healthy gait). The HRVT and HRAP are calculated by dividing the even harmonics (summed amplitudes of the first 20 even harmonics) by the odd harmonics (summed amplitudes of the first 20 odd harmonics), with higher HRs being characteristic of healthy, stable gait. Conversely, the ML accelerations exhibit one acceleration peak per stride, resulting in dominance of the first harmonic and subsequent odd harmonics. Here, the odd harmonics are in-phase and even harmonics are out-of-phase. Therefore, the HRML is calculated from a ratio of the odd harmonics divided by the even harmonics.

3. Results

Table 3 shows the results of gait parameters estimation across the two tested groups. Statistically significant differences, obtained by applying an independent samples *t*-test, are marked with asterisks. In Table 4, the correlation matrix computed for the gait parameters from the high fall risk group is also reported.

Table 3. Gait parameters results for the two groups.

	Patient Fall-Risk Level	
	Low-Risk	High-Risk
Walked distance (D), m	281 ± 100	183 ± 58 ***
Cadence, spm	103 ± 13	90 ± 15 ***
Stride time (T_{stride}), s	1.19 ± 0.18	1.37 ± 0.23 **
Stride time variability (COV), %	6.7 ± 3.1	8.5 ± 3.9
RMS (vertical acceleration), m/s^2	1.68 ± 0.59	1.05 ± 0.41 ***
HR$_{ML}$	1.66 ± 0.36	1.70 ± 0.42
HR$_{VT}$	1.93 ± 0.67	1.65 ± 0.56
HR$_{AP}$	1.69 ± 0.64	1.46 ± 0.38

***: $p < 0.001$ **: $p < 0.01$.

Table 4. Correlation matrix computed for the gait parameters from the high fall risk group. In bold: the strongest correlations among gait parameters.

	D	T_{stride}	COV	RMS	HR$_{ML}$	HR$_{VT}$	HR$_{AP}$
D	1.00						
T_{stride}	−0.28	1.00					
COV	0.56 **	0.21	1.00				
RMS	0.73 **	−0.32	−0.59 **	1.00			
HR$_{ML}$	−0.10	−0.55 **	−0.13	−0.021	1.00		
HR$_{VT}$	0.48 *	−0.28	−0.63 **	0.70 **	0.27	1.00	
HR$_{AP}$	0.42 *	−0.27	−0.59 **	0.64 **	0.35	0.88 **	1.00

*: $p < 0.05$ level (2-tailed) **: $p < 0.01$ level (2-tailed).

4. Discussion

The main results achieved during the preliminary experimental tests reported in this study can be summarized in the two following points. First, high fall-risk patients' gait was slower (0.51 m/s vs. 0.78 m/s on average), more variable (according to the estimated stride time variability), and less stable (the AP and VT directions, i.e., frontal stability) than normal gait (normative values taken from available literature). Statistically significant differences between the two groups were obtained for walked distance, stride time, cadence, and vertical acceleration RMS. In general terms, and despite the limitations of comparing these results with studies based on different protocols and selection criteria for the subjects, this is in accordance with previous studies that identified statistically significant reduced stride length and walking speed in high fall risk patients [26]. In our solution, the differences observed for stride time variability and harmonic ratios is not as high as it was observed to be in previous studies [28,34]; we believe that this could be an effect of the limited pool of tested subjects in this preliminary evaluation of the device.

Secondly, the results of the correlation analysis show the presence of a remarkably strong relationship between walked distance, stride time variability, and frontal stability, in the sense that a fast gait is likely to be less variable and more frontally stable than a slower pathologic gait. Such results confirmed the expectations and previous findings concerning gait subjects at high risk of falls, and in our view, confirmed the applicability of the proposed solution to the specific scenario of fall risk assessment in the elderly.

5. Conclusions

The proposed technological solution makes up part of the integrated solution for fall detection and fall risk assessment and mitigation as proposed in the I-DONT-FALL EU-funded project. The WIMU fall detector accomplished the aim of monitoring falls during daily life, allowing a full day's battery life and embedding custom-made fall detection algorithms. Moreover, the fall detector allowed the logging of fall data, which is useful for updating the existing fall detection computational solutions using real falls data. The designed WIMU data logger consisted of a fully functional logging unit that accurately and reliably logged data and implemented gait parameter estimation, embedding the processing in the unit.

Finally, it is worth noting that, the proposed solution for fall risk assessment and fall detection is fully integrated in an ICT solution aiming at a multi-factorial approach to risk assessment and at prompting intervention protocol in case of fall. Such a solution, the final result of the I-DONT-FALL project, achieved a commercially-ready integrated framework for both monitoring the elderly at home and reducing their fall risk by prescribing physical and cognitive activities tailored to the user.

Author Contributions: Conceptualization, V.G., A.M. and A.M.S.; methodology, V.G., A.M and A.M.S.; software, V.G.; validation, V.G., A.M. and A.M.S.; formal analysis, V.G., A.M., M.G. and A.M.S.; investigation, V.G., A.M., M.G. and A.M.S.; resources, V.G., A.M., M.G. and A.M.S.; data curation, V.G. and A.M.S.; writing—original draft preparation, V.G.; writing—review and editing, A.M., M.G. and A.M.S.; visualization, V.G.; supervision, A.M.S.; project administration, A.M.S. and V.G.; funding acquisition, A.M.S.

Funding: This research was funded by the EC, grant number CIP-297225 and by funds from the Italian Ministry of University and Research (MIUR).

Acknowledgments: Authors are pleased to recognize valuable advice and support throughout the I-DONT-FALL project-work, in particular the colleagues Dr. Matteo Melideo, project coordinator, Dr. Stelios Pantelopoulos, responsible of technical integration of the overall platform, and Drs. Roberta Annichiarico, responsible for the design of the clinical protocols adopted by the clinical teams involved in the project.

Conflicts of Interest: The authors declare no conflict of interest

References

1. Bergen, G.; Stevens, M.R.; Burns, E.R. Falls and fall injuries among adults aged ≥ 65 years—United States 2014. *MMWR Morb Mortal Wkly Rep* **2016**, *65*, 993–998. [CrossRef] [PubMed]
2. Fasano, A.; Plotnik, M.; Bove, F.; Berardelli, A. The neurobiology of falls. *Neurol. Sci.* **2012**, *33*, 1215–1223. [CrossRef] [PubMed]
3. Friedman, S.M.; Munoz, B.; West, S.K.; Rubin, G.S.; Fried, L.P. Falls and fear of falling: Which comes first? A longitudinal prediction model suggests strategies for primary and secondary prevention. *J. Am. Geriatr. Soc.* **2002**, *50*, 1329–1335. [CrossRef] [PubMed]
4. Lannin, N.A.; Clemenson, L.; McCluskey, A.; Lin, C.-W.C.; Cameron, I.D.; Barras, S. Feasibility and results of randomised pilot-study of pre-discharge occupational therapy home visits. *BMC Health Serv. Res.* **2007**, *7*, 42. [CrossRef] [PubMed]
5. Kraskowsky, L.H.; Finlayson, M. Factors affecting older adults' use of adaptive equipment: Review of the literature. *Am. J. Occup. Ther.* **2001**, *55*, 303–310. [CrossRef] [PubMed]

6. Hammond, A. What is the role of the occupational therapist? *Best Pract. Res. Clin. Rheumatol.* **2004**, *18*, 491–505. [CrossRef] [PubMed]
7. Ambrose, A.F.; Paul, G.; Hausdorff, J.M. Risk factors for falls among older adults: A review of the literature. *Maturitas* **2013**, *75*, 51–61. [CrossRef] [PubMed]
8. Cummings, S.R.; Nevitt, M.C.; Kidd, S. Forgetting falls: The limited accuracy of recall of falls in the elderly. *J. Am. Geriatr. Soc.* **1988**, *36*, 613–616. [CrossRef]
9. Menz, H.B.; Lord, S.R.; Fitzpatrick, R.C. Acceleration patterns of the head and pelvis when walking on level and irregular surfaces. *Gait Posture* **2003**, *18*, 35–46. [CrossRef]
10. Moe-Nillsen, R.; Aaslund, M.C.; Hodt-Billington, C.; Helbostad, J.L. Gait variability measures may represent different constructs. *Gait Posture* **2010**, *32*, 98–101. [CrossRef] [PubMed]
11. Silva, P.A.; Nunes, F.; Vasconcelos, A.; Kerwin, M.; Moutinho, R.; Teixeira, P. Using the smartphone accelerometer to monitor fall risk while playing a game: The design and usability evaluation of dance! Don't fall. In *Foundations of Augmented Cognition*; Springer: Berlin, Germany, 2013; pp. 754–763.
12. Garcia, J.A.; Pisan, Y.; Tan, T.C.; Navarro, K.F. Assessing the Kinect's capabilities to perform a time-based clinical test for fall risk assessment in older people. In *Entertainment Computing–ICEC 2014*; Springer: Berlin, Germany, 2014; pp. 100–107.
13. Taylor, M.J.; Shawis, T.; Impson, R.; Ewins, K.; McCormick, D.; Griffin, M. Nintendo Wii as a training tool in falls prevention rehabilitation: Case studies. *J. Am. Geriatr. Soc.* **2012**, *60*, 1781–1783. [CrossRef] [PubMed]
14. Williams, M.A.; Soiza, R.L.; Jenkinson, A.M.; Stewart, A. Exercising with Computers in Later Life (EXCELL)-pilot and feasibility study of the acceptability of the Nintendo WiiFit in community-dwelling fallers. *BMC Res. Notes* **2010**, *3*, 238. [CrossRef] [PubMed]
15. Abbate, S.; Avvenuti, M.; Bonatesta, F.; Cola, G.; Corsini, P.; Vecchio, A. A smartphone-based fall detection system. *Pervasive Mob. Comput.* **2012**, *8*, 883–899. [CrossRef]
16. Yu, M.; Rhuma, A.; Naqvi, S.M.; Wang, L.; Chambers, J. A posture recognition-based fall detection system for monitoring an elderly person in a smart home environment. *IEEE Trans. Inform. Technol. Biomed.* **2012**, *16*, 1274–1286.
17. Fortino, G.; Gravina, R. Fall-MobileGuard: A smart real-time fall detection system. In Proceedings of the 10th EAI International Conference on Body Area Networks, Sydney, Australia, 28–30 September 2015; pp. 44–50.
18. Chen, J.; Kwong, K.; Chang, D.; Luk, J.; Bajcsy, R. Wearable sensors for reliable fall detection. In Proceedings of the 2005 IEEE Engineering in Medicine and Biology 27th Annual Conference, Shanghai, China, 17–18 January 2006; pp. 3551–3554.
19. Doughty, K.; Lews, R.; Mcintosh, A. The design of practical and reliable fall detector for community and institutional telecare. *J. Telemed. Telecare* **2000**, *6*, 150–154. [CrossRef]
20. Bianchi, F.; Redmond, S.J.; Narayanan, M.R.; Cerutti, S.; Lovell, N.H. Barometric pressure and triaxial accelerometry-based falls event detection. *IEEE Trans. Neural Syst. Rehabil. Eng.* **2010**, *18*, 619–627. [CrossRef] [PubMed]
21. Wu, G.; Xue, S. Portable preimpact fall detector with inertial sensor. *IEEE Trans. Neural Syst. Rehabil. Eng.* **2008**, *16*, 178–183. [PubMed]
22. Hamm, J.; Money, A.G.; Atwal, A.; Paraskevopoulos, I. Fall prevention intervention technologies: A conceptual framework and survey of the state of the art. *J Biomech Inform.* **2016**, *59*, 319–345. [CrossRef] [PubMed]
23. Sabatini, A.M.; Ligorio, G.; Mannini, A.; Genovese, V.; Pinna, L. Prior-to-and post-impact fall detection using inertial and barometric altimeter measurements. *IEEE Trans. Neural Syst. Rehabil. Eng.* **2016**, *24*, 774–783. [CrossRef] [PubMed]
24. Bair, W.; Prettyman, M.G.; Beamer, B.A.; Rogers, M.W. Kinematic and behavioral analyses of protective stepping strategies and risk for falls among community living older adults. *Clin. Biomech.* **2016**, *36*, 74–82. [CrossRef] [PubMed]
25. Zijlstra, W.; Hof, A.L. Assessment of spatio-temporal gait parameters from trunk accelerations during human walking. *Gait Posture* **2003**, *18*, 1–10. [CrossRef]
26. Taylor, M.E.; Delbaere, K.; Mikolaizak, A.S.; Lord, S.R.; Close, J.C.T. Gait parameter risk factors for falls under simple and dual task conditions in cognitively impaired older people. *Gait Posture* **2013**, *37*, 126–130. [CrossRef] [PubMed]

27. Gillain, S.; Boutaayamou, M.; Beaudart, C.; Demonceau, M.; Bruyère, O.; Reginster, J.Y.; Garraux, G.; Petermans, J. Assessing gait parameters with accelerometer-based methods to identify older adults at risk of falls: A systematic review. *Eur. Geriatr. Med.* **2018**, *9*, 435–448. [CrossRef]
28. Hausdorff, J.M.; Eldelberg, H.K.; Mitchell, S.L.; Goldberger, A.L.; Wei, J.Y. Increased gait unsteadiness in community-dwelling elderly fallers. *Arch. Phys. Med. Rehab.* **1997**, *78*, 278–283. [CrossRef]
29. IDONTFALL Project. Available online: http://www.idontfall.eu/ (accessed on 29 September 2018).
30. Genovese, V.; Mannini, A.; Sabatini, A.M. IDONTFALL Deliverable D3.1: Detailed Technical Specification of Fall Detection and Prevention Services. 2012. Available online: http://www.idontfall.eu/sites/default/files/deliverables (accessed on 29 September 2018).
31. Barban, F.; Annicchiarico, R.; Melideo, M.; Federici, A.; Lombardi, M.G.; Giuli, S.; Ricci, C.; Adriano, F.; Griffini, I.; Silvestri, F.; et al. Reducing fall risk with combined motor and cognitive training in elderly fallers. *Brain Sci.* **2017**, *7*, 19. [CrossRef] [PubMed]
32. Bourke, A.K.; O'Donovan, K.J.; O'Laighin, G. The identification of vertical velocity profiles using an inertial sensor to investigate pre-impact detection of falls. *Med. Eng. Phys.* **2008**, *30*, 937–946. [CrossRef] [PubMed]
33. Bellanca, J.L.; Lowry, K.A.; VanSwearingen, J.M.; Brach, J.S.; Redfern, M.S. Harmonic ratios: A quantification of step to step symmetry. *J. Biomech.* **2013**, *46*, 828–831. [CrossRef] [PubMed]
34. Menz, H.B.; Lord, S.R.; Fitzpatrick, R.C. Acceleration patterns of the head and pelvis when walking are associated with risk of falling in community-dwelling older people. *J. Gerontol. (Series A: Biol Sci Med Sci)* **2003**, *58*, M446–M452. [CrossRef]

technologies

Article

Determining the Reliability of Several Consumer-Based Physical Activity Monitors

Joshua M. Bock [1,2], Leonard A. Kaminsky [1,3], Matthew P. Harber [1] and Alexander H. K. Montoye [1,4,*]

[1] Clinical Exercise Physiology Program, Ball State University, Muncie, IN 47306, USA; jmbock@healthcare.uiowa.edu (J.M.B.); kaminskyla@bsu.edu (L.A.K.); mharber@bsu.edu (M.P.H.)
[2] Department of Physical Therapy and Rehabilitation Science, University of Iowa, Iowa City, IA 52242, USA
[3] Fisher Institute of Health and Well-Being, Ball State University, Muncie, IN 47306, USA
[4] Department of Integrative Physiology and Health Science, Alma College, Alma, MI 48801, USA
* Correspondence: montoyeah@alma.edu; Tel.: +1-989-463-7923

Received: 31 May 2017; Accepted: 21 July 2017; Published: 24 July 2017

Abstract: Limited research exists on the reliability of consumer-based physical activity monitors (CPAMs) despite numerous studies on their validity. Consumers often purchase CPAMs to assess their physical activity (PA) habits over time, emphasizing CPAM reliability more so than their validity; therefore, the purpose of this study was to investigate the reliability of several CPAMs. In this study, 30 participants wore a pair of four CPAM models (Fitbit One, Zip, Flex, and Jawbone Up24) for a total of eight monitors, while completing seven activities in the laboratory. Activities were completed in two consecutive five-minute bouts. Participants then wore either all wrist- or hip-mounted CPAMs in a free-living setting for the remainder of the day. Intra-monitor reliability for steps (0.88–0.99) was higher than kcals (0.77–0.94), and was higher for hip-worn CPAMs than for wrist-worn CPAMs ($p < 0.001$ for both). Inter-monitor reliability in the laboratory for steps (0.81–0.99) was higher than kcals (0.64–0.91) and higher for hip-worn CPAMs than for wrist-worn CPAMs ($p < 0.001$ for both). Free-living correlations were 0.61–0.98, 0.35–0.96, and 0.97–0.98 for steps, kcals, and active minutes, respectively. These findings illustrate that all CPAMs assessed yield reliable estimations of PA. Additionally, all CPAMs tested can provide reliable estimations of physical activity within the laboratory but appear less reliable in a free-living setting.

Keywords: physical activity; accelerometry; steps; energy expenditure; activity tracker

1. Introduction

Despite the well-known benefits of regularly engaging in physical activity (PA), half, or more, of U.S. adults do not meet the 2008 Physical Activity Guidelines for Americans [1–3]. To better understand the role of PA in improving health and reducing disease burden, it is important to measure PA accurately and reliably. High-quality measurement techniques allow researchers to identify which activity intensities provide optimal health benefits, monitor intermittent bouts of PA, and more accurately assess the effectiveness of interventions for promoting behavior change [4]. Consumer-based PA monitors (CPAMs) are common accessories with one in ten adults in the United States owning a CPAM [5]. During the first fiscal quarter of 2016, 19.7 million fitness trackers were sold worldwide; a 67.2% increase from quarter one of 2015. Fitbit Inc. was the largest distributor of fitness trackers during quarter one of 2015 and 2016 with 32.6 and 24.5% market share, respectively [6]. Despite surging popularity of these devices, one in three consumers who purchase a CPAM stops using it after six months [5]. The reasons for the high dropout in using CPAMs are not well understood, but they may be partly related to a lack of understanding on how well (e.g., accurately and reliably) the CPAMs capture PA levels and patterns over time.

While studies of device accuracy are common, much less research has investigated CPAM reliability [7]. For instance, only one study has assessed intra-monitor reliability (e.g., test-retest reliability). Kooiman et al. [8] assessed the intra-monitor reliability of the Fitbit Flex (Fitbit Inc., San Francisco, CA), Jawbone Up (AliphCom dba Jawbone, San Francisco, CA, USA), and Fitbit Zip (Fitbit Inc., San Francisco, CA, USA) to estimate steps using two bouts on a treadmill at 3.0 mph for 30 min. High intra-class correlations (ICCs) were found (0.81–0.90, for the Fitbit Flex, Fitbit Zip, and Jawbone Up), but these results are limited to a single activity and did not assess other variables, such as kcals or active minutes [8].

Four studies have assessed inter-monitor reliability (agreement between various monitors used during the same assessment). These studies ranged from case studies to those with 30 participants and assessed the Fitbit Ultra, Fitbit One, and Fitbit Flex. All four studies found Pearson correlations >0.90 for both steps and kcal measurements during ambulation in laboratory settings [9–11] or across free-living settings [12].

Current CPAM reliability research is limited regarding the diversity of activities tested (mainly walking and jogging) and the variables assessed (mainly steps). Additionally, several studies have evaluated inter-monitor reliability exclusively despite intra-monitor reliability being more relevant to assessing PA habits over time as consumers rarely use multiple CPAMs at a given time. Furthermore, little work has been done to assess CPAM inter-monitor reliability in a free-living setting. The inclusion of multiple settings is critical as several studies have reported setting-oriented differences in CPAM performance [13,14]. This study's purpose was to assess the intra- and inter-monitor reliability of several CPAMs for steps and kcals during a variety of activities, as well as the inter-monitor reliability to estimate steps, kcals, and active minutes in a free-living setting.

2. Methods

Participants

In this study, 30 (9M/21F) young adults were recruited from the East-Central region of Indiana. To be eligible for this study, participants had to be free of gait abnormalities, free of acute illness, between the ages of 18 and 80 years, not pregnant, and capable of completing the protocol without undue fatigue.

Prior to participating in the study all participants provided written informed consent approved by Ball State University's Institutional Review Board. All participants were right-handed and Caucasian; demographic information is shown in Table 1.

Table 1. Demographic information on participants categorized per analysis.

	All Participants	ICCs ($n = 28$)	Pearson ($n = 30$)	FL Hip ($n = 15$)	FL Wrist ($n = 15$)
Age (years)	23.1 ± 2.1	23.0 ± 2.1	23.0 ± 2.0	23.8 ± 2.4	22.4 ± 1.7
BMI (kg·m^{-2})	23.3 ± 3.4	23.4 ± 3.5	23.2 ± 3.3	23.3 ± 2.7	23.3 ± 4.0
Treadmill Brisk (km·h^{-1})	5.3 ± 0.3	5.3 ± 0.3	5.5 ± 0.3	- -	- -
Treadmill Jog (km·h^{-1})	8.7 ± 2.1	8.9 ± 2.1	8.9 ± 2.1	- -	- -

Kcal = kilocalories. BMI = body mass index. ICCs = data from participants used during intra-monitor analysis. Pearson = data from participants used during inter-monitor analysis. FL Hip = data from participants assigned hip-worn monitors during free-living portion of study. FL Wrist = data from participants assigned wrist-worn monitors during free-living portion of study. Data presented as mean ± standard deviation.

3. Equipments

During the laboratory visit, participants wore eight CPAMs (one pair of four different models). Descriptions of the CPAMs used are provided below.

Fitbit One (FO; Fitbit Inc., San Francisco, CA, USA): The FO, a hip-worn CPAM weighing 8.5 grams was used to estimate steps and kcals in the laboratory setting, as well as steps, kcals, and active minutes during the free-living portion of the study. Data are quantified by the FO by utilizing the demographic information entered into the monitor, as well as through measurements made via accelerometer hardware within the monitor. This CPAM has an internal, rechargeable battery and provides real-time feedback to its user. The FO has the capability to synchronize with the Fitbit Mobile Application via a Bluetooth connection allowing the user to track PA over time. Data from the FO were collected from the Fitbit Mobile Application before and after each activity.

Fitbit Zip (Fitbit Inc., San Francisco, CA, USA): The FZ, is a hip-worn CPAM weighing 8.5 grams and was used to estimate steps and kcals in the laboratory setting as well as steps, kcals, and active minutes during the free-living portion of the study. Data are quantified by the FZ by utilizing the demographical information entered into the monitor, as well as through measurements made via the accelerometer hardware within the monitor. The FZ uses a CR-2032 watch battery and has the capability to synchronize with the Fitbit Mobile Application via a Bluetooth connection. Data from the FZ were collected from the device's built-in display screen before and after each activity.

Jawbone Up24 (AliphCom dba Jawbone, San Francisco, CA, USA): The JU, a wrist-worn CPAM weighing 22.7 grams and was used to estimate steps and kcals in the laboratory setting, as well as steps, kcals, and active minutes during the free-living portion of the study. Data are quantified by the JU by utilizing the demographical information entered into the monitor, as well as through measurements made via accelerometer hardware within the monitor. This CPAM utilizes an internal, rechargeable battery and can provide real-time feedback to its user via Bluetooth connection and the UP Mobile Application. Data from the JU were collected from the UP Mobile Application before and after each activity.

Fitbit Flex (Fitbit Inc., San Francisco, CA, USA): The FF, a wrist-worn CPAM weighing 17.0 grams and was used to estimate steps and kcals in the laboratory setting, as well as steps, kcals, and active minutes during the free-living portion of the study. Data are quantified by the FF by utilizing the demographical information entered into the monitor, as well as through measurements made via accelerometer hardware within the monitor. This monitor utilizes an internalized, rechargeable battery and requires the Fitbit Mobile Application and a Bluetooth connection to track PA. Data from the FF were collected from the Fitbit Mobile Application before and after each activity.

4. Protocol

Participants came to the Clinical Exercise Physiology Laboratory at Ball State University twice. During visit 1, participants completed an informed consent and had their height and weight measurements taken via scale (to the nearest 0.1 kg) and stadiometer (to the nearest 1.0 cm), which were then entered into each CPAM's respective mobile applications in addition to age, sex, and hand dominance. Researchers then fitted the CPAMs to the participants; initial readings of steps and kcals were collected from all CPAMs while the participant was in a seated position.

Following baseline data collection, participants completed a laboratory-based activity protocol. Each participant underwent an identical protocol where all activities lasted for five minutes, excluding transition time between activities. The only exception was the 'climbing stairs' activity in which all participants ascended and descended a flight of stairs five times at a self-selected pace. All activities were performed twice in succession with CPAM data collected before and after each activity bout to allow for the intra-monitor reliability analysis. It should be noted that data collected from the CPAMs were done so in the same order (FO, FZ, FF, JU) to minimize variability. Additionally, transition times between activity bouts lasted approximately one to three minutes and were determined by CPAM synchronization rate following activity bouts. The activity protocol was structured in the following order: typing, reading, sweeping (participants swept confetti into a pile within a ~10 m^2 section of the laboratory), slow treadmill walk at 3.2 km/h, brisk treadmill walk (4.8–5.6 km/h), treadmill jog (6.4–12.9 km/h), and ascending/descending stairs. Participants chose paces for the brisk treadmill walk, treadmill jog, and stairs activities.

5. Data Cleaning and Analysis

Intra-monitor reliability was assessed via intra-class correlations (ICCs) independently for each CPAM model (FO, FZ, JU, and FF) and outcome variable of interest (steps and kcals). Data used in this analysis came from a single monitor of each brand during both bouts of each activity. For wrist-worn CPAMs, data from the distal monitor were used, whereas data for hip-worn CPAMs came from the anterior monitor. Inter-monitor reliability was assessed by comparing data from each monitor brand with its pair (e.g., one FF against the other FF) for the first activity bout exclusively. Pearson correlations, calculated for each CPAM model and outcome variable of interest, were used to define the inter-monitor reliability for each CPAM. Both intra- and inter-monitor reliability analyses used protocol-wide data. That is, for each participant, there was a single ICC and Pearson correlation calculated using data from all activities, for a total of 30 data points for each analysis points. It should be noted that these analyses occurred after exclusion criteria were applied (see below).

Participants also completed a free-living protocol after their first laboratory visit. During this protocol, the participants continued to wear either hip-worn (FOs and FZs) or wrist-worn (FFs and JUs) CPAMs. Participants were assigned either hip- or wrist-worn CPAMs as the researchers presumed wearing all eight CPAMs for most of a day would be uncomfortable for participants and may, therefore, alter their behavior and/or reduce compliance with wearing the devices. These CPAMs were worn for the remainder of the day then returned to the lab the following morning (visit two) when the research staff collected CPAM monitors and data concluding participants' involvement in the study. The CPAMs assigned to the participants were arranged among participants so that each placement site (hip or wrist) was used by 15 participants. Free-living data were analyzed using Pearson correlations in a similar fashion to the laboratory inter-monitor reliability analysis.

A pair of exclusion criteria was applied to the collected CPAM data to remove data likely influenced by monitor malfunctions. The exclusion criteria for laboratory data were (1) data were negative (e.g., steps decreased following an activity) or (2) the kcals variable was not updated for a given CPAM following an activity. Exclusion criteria for the free-living portion were (1) data were negative or (2) steps taken over the remainder of the day were ≤150 steps. Once these criteria were applied, a repeated-measures analysis of variance (RM-ANOVA) with Tukey's post-hoc was used to determine if significant differences existed among the ICCs. Bland-Altman plots were created using step and kcal data from both bouts (intra) and monitor pairs (inter) to illustrate the nature of CPAM differences.

Additionally, median absolute differences (MAD) have been used alongside correlations to characterize agreement, as done in previous work [15]. Initially, absolute differences were calculated for each monitor per participant and activity. Then, the medians of the absolute differences were determined per monitor and participant and presented as MAD. Median percent difference (MPD) was calculated in a similar fashion using percent differences in place of absolute differences. MAD and MPD were calculated using step and kcal data from both bouts (intra-monitor) and each pair of monitors (inter-monitor).

All analyses were conducted in SPSS version 23.0 (IBM, Armonk, NY, USA) and Microsoft Excel (Microsoft, Redmond, WA). Statistical significance was defined a priori as $\alpha < 0.05$. Nomenclature for correlation strength was designated as follow: high ($r = 0.80–1.00$), moderately high ($r = 0.60–0.79$), low ($r = 0.40–0.59$), or no relationship ($r = 0.00–0.19$) as set forth by Safrit et al. [16].

6. Results

6.1. Intra-Monitor Reliability

Two participants were excluded from the intra-monitor reliability analysis due to errors encountered during data collection (e.g., poor synchronization of the mobile application), resulting in 28 participants' data being used during analysis. Additionally, exclusion criteria removed 11.8% of step and 8.3% of kcal data from participants included in the analysis. The ICCs for steps and kcals are shown in Figure 1. All step ICCs were high ($\geq$0.80) with the FZ significantly higher than the JU and FF ($p < 0.05$). All ICCs for kcals were moderately high (0.60–0.79) with the FO having significantly higher reliability than the JU and FF ($p < 0.05$). Data for MAD and MPD are shown in Table 2. Recorded estimations per CPAM and activity from the first visit are shown in Table 3. Intra-monitor MAD and MPD step values for the hip-worn CPAMs were significantly lower (better) than the wrist-worn CPAMs ($p < 0.01$) without a concurrent difference in kcal estimations ($p = 0.46$ and 0.53, respectively). The JU had the largest average MAD for steps (11), kcals (2.1), and the highest MPD for kcals (13.9%); the FF had the largest MPD for steps (7.2%). Figure 2 (steps) and Figure 3 (kcals) illustrate that CPAM error was higher in some cases during activities with higher predicted PA; however, these results may be partly influenced by outliers as they were not excluded from analysis. In general, the 95% limits of agreement were narrower for the hip-worn CPAMs compared to the wrist-worn CPAMs.

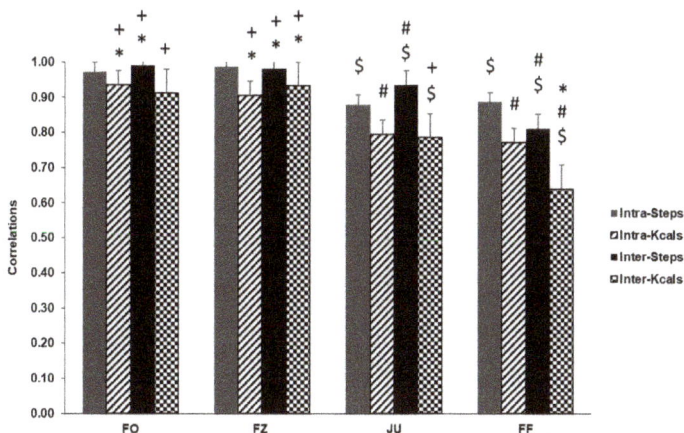

Figure 1. Intra-class correlations (Intra-) and Pearson correlations (inter-) for laboratory data. FO = Fitbit One. FZ = Fitbit Zip. JU = Jawbone Up24. FF = Fitbit Flex. Intra-Steps = intra-class coefficient for steps. Intra-kcals = intra-class coefficient for Calories. Inter-Steps = Pearson correlations for steps. Inter-Kcals = Pearson correlation for Calories. # statistically different from FO. $ statistically different from FZ. * statistically different from JU. + statistically different from FF. Statistical significance was defined as $p < 0.05$ for all.

Table 2. Median absolute differences and median absolute percent differences across entire laboratory protocol.

	FO	FZ	JU	FF
Intra-monitor reliability				
Steps	1.9 + (0.4) *	3.3 + (0.7) *	7.5 (2.3) #,$	10.9 #,$ (2.6)
Kcals	1.5 (8.8)	1.5 (9.1)	1.8 (12.5)	1.5 (8.9)
Inter-monitor reliability (Lab)				
Steps	0.5 *,+ (0.1) *,+	0.5 (0.1) +	2.8 # (0.7) #	4.0 # (1.4) #,$
Kcals	0.5 *,+ (4.6) *,+	0.5 *,+ (5.9) *,+	1.3 #,$ (9.4) #,$	1.5 #,$ (9.1) #,$
Inter-monitor reliability (FL)				
Steps	35 (2.1)	128 (7.0)	731 (8.1)	154 (5.2)
Kcals	34 (5.1)	60 (8.3)	26 (4.1)	88 (11.5)
Active Minutes	0 (0.0)	0 (0.0)	6 (8.5)	0 (0.0)

Kcals = kilocalories. Data presented as MAD (MPD). MAD = median absolute difference. MPD = median absolute percent difference. FL = free-living. FO = Fitbit One. FZ = Fitbit Zip. JU = Jawbone Up24. FF = Fitbit Flex. # significantly different from FO ($p < 0.05$). $ significantly different from FZ. * significantly different from JU. + significantly different from FF.

Table 3. Physical activity estimations per monitor and activity from visit one.

Steps	Typing	Reading	Sweeping	Slow TM	Brisk TM	TM Jog	Stairs
FO	0 ± 1	0 ± 0	10 ± 43	476 ± 35	587 ± 31	725 ± 105	113 ± 21
FZ	2 ± 8	1 ± 3	1 ± 3	461 ± 60	583 ± 100	744 ± 105	123 ± 36
JU	2 ± 6	0 ± 2	266 ± 166	426 ± 74	575 ± 127	779 ± 136	125 ± 46
FF	1 ± 3	3 ± 8	327 ± 116	399 ± 142	529 ± 151	757 ± 185	153 ± 128
Kcals							
FO	9.2 ± 2.0	7.7 ± 1.4	10.1 ± 2.6	20.8 ± 4.4	32.7 ± 6.4	48.9 ± 12.3	13.2 ± 3.7
FZ	11.6 ± 8.7	7.7 ± 1.2	7.8 ± 2.0	41.3 ± 25.8	41.3 ± 6.7	56.0 ± 9.0	12.1 ± 4.4
JU	10.7 ± 3.9	7.6 ± 1.7	17.6 ± 7.1	21.5 ± 7.4	31.1 ± 9.6	56.2 ± 22.4	12.1 ± 11.3
FF	12.8 ± 15.6	8.1 ± 1.7	28.9 ± 9.5	33.0 ± 11.7	39.1 ± 11.9	58.7 ± 18.3	14.5 ± 7.8

TM = treadmill. Kcal = kilocalories. FO = Fitbit One. FZ = Fitbit Zip. JU = Jawbone Up24. FF = Fitbit Flex. Data presented as mean ± standard deviation.

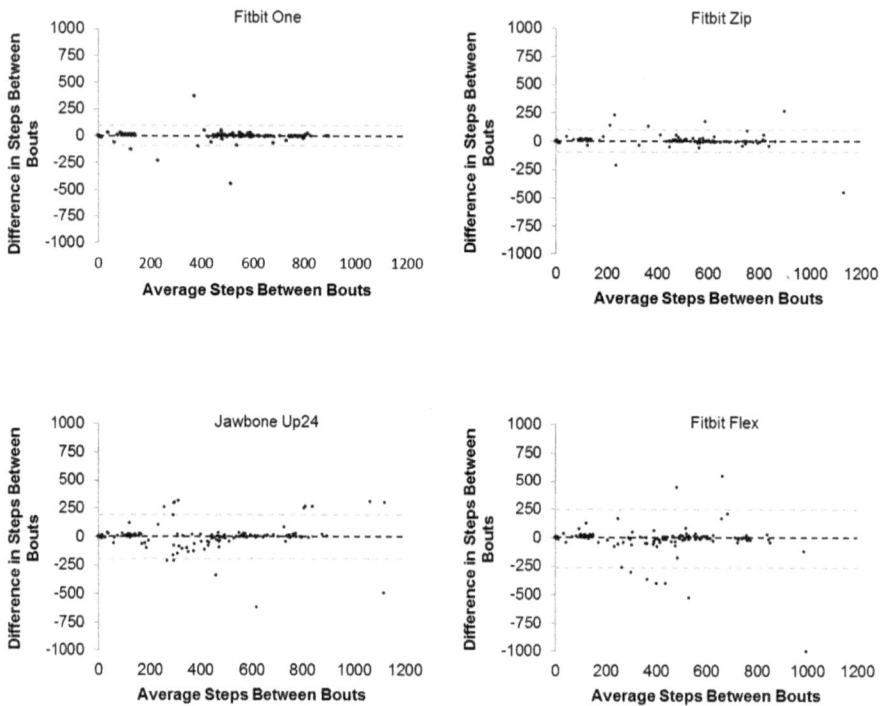

Figure 2. Bland-Altman plots with 95% limits of agreement calculated using the intra-monitor step data from all activities completed by each participant.

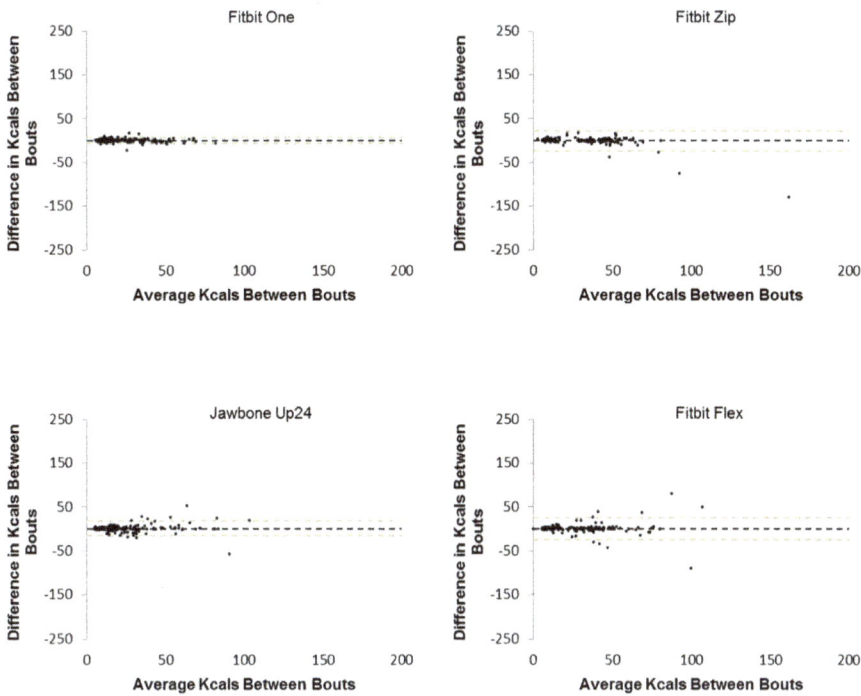

Figure 3. Bland-Altman plots with 95% limits of agreement calculated using the intra-monitor kilocalorie (kcals) data from all activities completed by each participant.

6.2. Inter-Monitor Reliability: Laboratory Setting

All 30 participants' data were included in the inter-monitor reliability analysis. Correlations for steps and kcals are shown in Figure 1. Prior to analysis, 11.5% (step) and 8.2% (kcal) data were removed per exclusion criteria mentioned above. All step correlations were high ($\geq$0.80). Both hip-worn CPAMs (FO and FZ) had correlations significantly higher than the wrist-worn CPAMs (JU and FF, $p < 0.05$). Kcal correlations for the FO and FZ were high ($\geq$0.80); the JU and FF correlations were moderately high (0.60–0.79). Correlations were significantly higher for the FO than the FF, the FZ than the JU and FF, and the JU than the FF ($p < 0.05$). Results from MAD and MPD are shown in Table 2. Recorded estimations per CPAM and activity from the first visit are shown in Table 3. Inter-MAD and MPD values were significantly lower in hip-worn CPAMs than wrist-worn CPAMs ($p < 0.05$ and < 0.01, respectively). The JU had the largest MAD and MPD for kcals (2.7 and 14.2%), the FF had the largest MAD value for steps (7), and the FZ had the largest MPD for steps (6.2%). For both steps (Figure 4) and kcals (Figure 5), the 95% limits of agreement were narrower for the hip-worn CPAMs compared to the wrist-worn CPAMs.

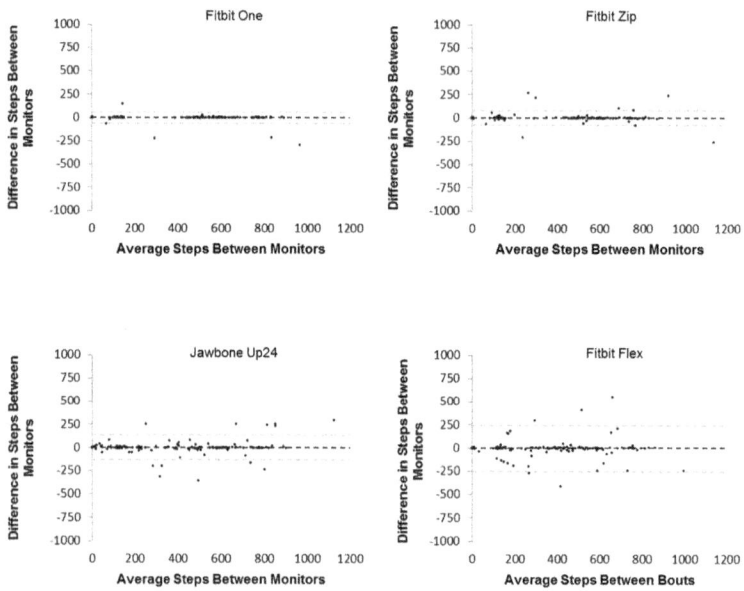

Figure 4. Bland-Altman plots with 95% limits of agreement created using the inter-monitor step data from all activities completed by each participant.

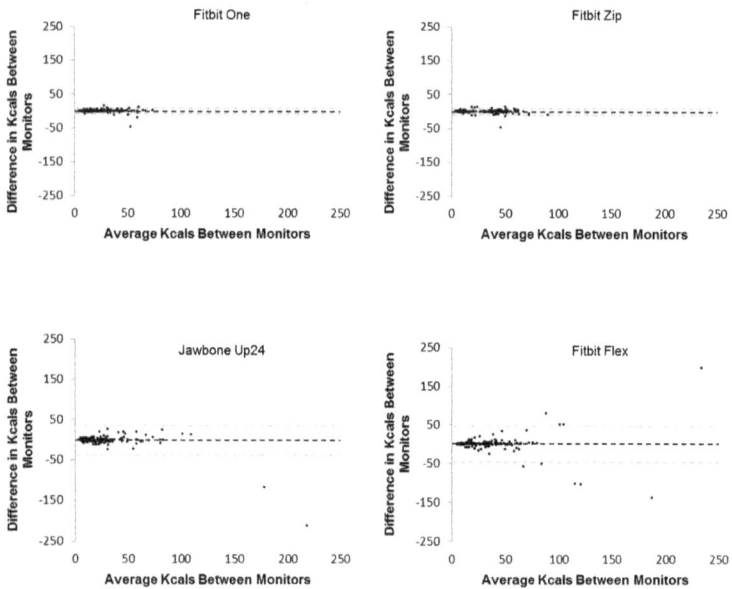

Figure 5. Bland-Altman plots with 95% limits of agreement created using the inter-monitor kilocalorie (kcals) data from all activities completed by each participant.

6.3. Inter-Monitor Reliability: Free-Living Setting

Each pair of CPAMs (wrist- or hip-worn) was worn by fifteen participants. A small percentage of step (3.6%), kcal (0.0%), and active minute (5.0%) were removed per exclusion criteria. Minimum wear time was not mandated; however, mean wear time was 5.7 ± 3.8 h. Correlations for steps, kcals, and active minutes for all CPAMs are shown in Figure 6. Most CPAMs had high inter-monitor reliability for all variables, except for kcals for the FO (low), active minutes for the FZ (moderate), and steps/kcals for the FF (moderately high). The abnormally low FO kcal and FZ active minutes correlations are attributable to infrequent outliers illustrated in Figures 7–9. MAD and MPD data paralleled data collected in the laboratory setting; that is, wrist-worn CPAMs displayed greater (worse) MAD and MPD data compared to the hip-worn CPAMs. JU had the highest step and active minute MAD and MPDs while the FF had the largest MAD and MPD for kcals.

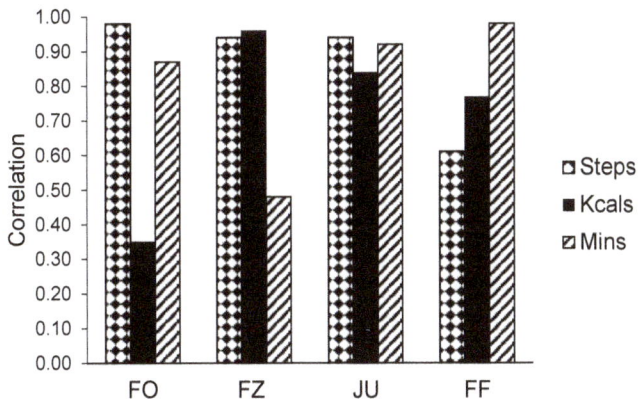

Figure 6. Pearson correlations (inter-monitor reliability) of the free-living data. Kcals = Calories. Mins = active minutes.

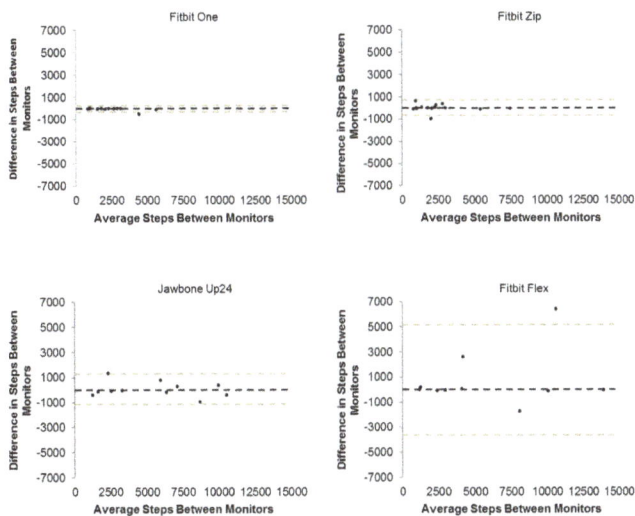

Figure 7. Bland-Altman plots with 95% limits of agreement created using the free-living steps data from each participant.

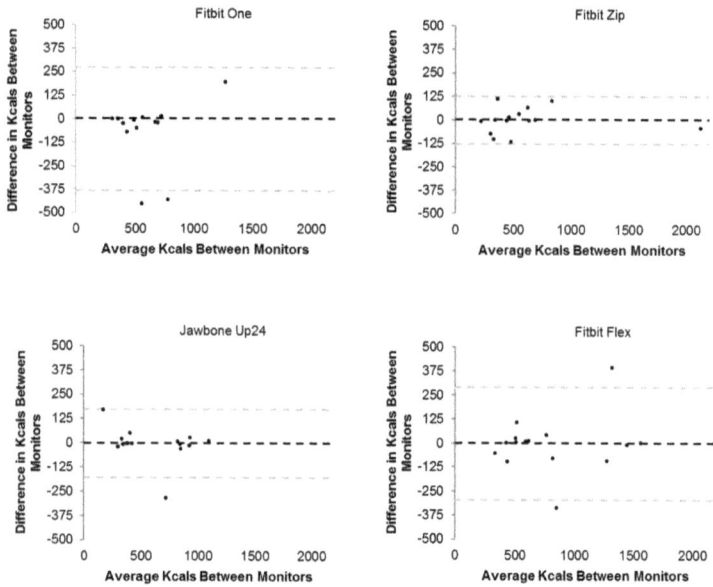

Figure 8. Bland-Altman plots with 95% limits of agreement created using the free-living kilocalories (kcals) data from each participant.

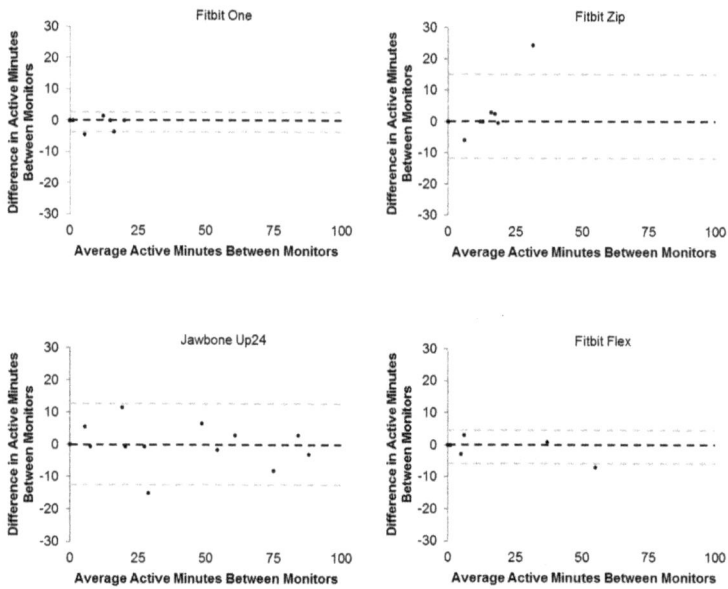

Figure 9. Bland-Altman plots with 95% limits of agreement created using the free-living active minutes data from each participant.

7. Discussion

This study found that all CPAMs had high intra-monitor reliability (≥0.80) for steps in a laboratory setting; however, the hip-worn CPAMs were significantly more reliable than the wrist-worn CPAMs. The ICCs in the present study are higher than those found by Kooiman et al. [8] who determined intra-monitor reliability for steps using the FF, JU, and FZ; their ICCs were 0.81, 0.83, and 0.90, respectively compared to 0.89, 0.88, and 0.99 in our study also using a laboratory setting [8]. Discrepancies between the studies could be attributable to differences in activity protocols. Kooiman et al. [8] used a single treadmill walking activity, whereas the present study used seven different activities, including both ambulatory (e.g., walking) and non-ambulatory (e.g., typing) tasks. The larger number and greater variety of activities used in our study builds upon preexisting CPAM literature and advances our understanding on how these devices perform during free-living activity. While no other studies have investigated the intra-monitor reliability of the FO for steps or any CPAMs to estimate kcals, we found lower reliability for kcal estimated than for step estimates, both in terms of lower correlations but also higher MPD. Our findings show consistently high intra-monitor reliability, especially for step estimates, with a variety of CPAMs and activities.

High correlations (≥0.80) were also observed in the inter-monitor reliability analyses for all CPAMs when estimating steps but only for the hip-worn CPAMs when estimating kcals; wrist-worn CPAMs had moderately-high correlations for kcal estimations. When examining CPAM validity, most studies show higher accuracy for step estimations than kcal estimates [7,16,17]. Therefore, available evidence suggests that step estimations from CPAMs are both more reliable and more valid than kcal estimations. The correlations obtained in the present study were comparable to those reported by Diaz et al. for steps (0.97 vs. 0.99) and kcals (0.94 vs. 0.97), as well as those of Takacs et al. [11] for steps (0.96 vs. 1.00), respectively [8,10]. It is important to note the consistently high correlations across various activity protocols indicating that reliability remains high even with the inclusion of a variety of activities, contrasting validity research where inclusion of diverse activities lowers CPAM validity [18].

CPAM correlation point estimates from the free-living portion of this study were comparable to, or lower than, those found during the laboratory portion. Most correlations were moderately high, although there were four instances when CPAM performance failed to meet the moderately high threshold. These instances included the FO for kcals, the FZ for active minutes, and the FF for steps and active minutes. The JU was the only CPAM whose correlations met the moderately high or greater criteria for all PA variables. A case study examining inter-monitor reliability of 10 Fitbit Ultra devices in an eight-day free-living trial found considerably higher reliability coefficients (0.995–1.000) for daily step counts than our study [12]. The Fitbit Ultra is a hip-worn CPAM, which partially explains the strong reliability found in their study. Additionally, only a portion of the day was spent in a free-living setting in our study, there was likely smaller variability in the data collected subsequently causing lower correlation coefficients than seen in the work of Dontje et al [12]. However, recent works have illustrated that CPAM's underestimate PA in free-living settings [14] and that the variability of these estimations is not consistent between CPAM models [13]. Collectively, available research suggests high or moderately-high reliability for most CPAMs and most dependent variables tested in free-living settings, supporting the use of these CPAMs during field-based PA monitoring [12].

While all CPAMs in the present study yielded moderately-high to high intra- and inter-monitor reliability in the laboratory, the hip-worn CPAMs (FO and FZ) had higher reliability than the wrist-worn CPAMs (JU and FF), both in terms of correlations as well as smaller (better) intra- and inter-monitor MAD and MPD values and generally narrower (better) 95% limits of agreement when examined using Bland-Altman plots. Given the greater variability and higher accelerations of arm movement compared to hip movement during basic tasks, these results were expected. However, wrist-worn activity monitors have better user compliance than hip-worn monitors [2,19,20]. Additionally, there are a greater number of wrist-worn CPAMs than hip-worn CPAMs on the market suggesting that wrist-worn CPAMs may be the more popular models. Accordingly, the choice of CPAM placement (wrist vs. hip) may depend on the importance of optimal reliability vs. optimal compliance and comfort.

All CPAMs in the present study collect and interpret PA data based upon accelerometer-based sensors within the device. More recently, manufacturers have produced CPAMs which incorporate variables, such as heart rate or other variables (e.g., skin temperature), into their algorithms (e.g., Apple Watch and Fitbit Charge). Indeed, a recent study showed these multi-sensor CPAMs showed improved energy expenditure estimations compared to single-sensor CPAMs [13]. As multi-sensor CPAMs become more common, the reliability of their newer variables (e.g., heart rate) and the influence of newer variables on other preexisting variables (e.g., kcals) should be investigated as there is likely crosstalk between sensors.

This study did not evaluate CPAM validity, but the relationship between CPAM validity and reliability is worth considering. A recent review article by Evenson et al. [7] reported results from over 20 validity and reliability studies, finding high validity and inter-monitor reliability for steps using treadmill-oriented protocols. Evenson also noted lower monitor validity during non-ambulatory activities and when the CPAMs were used in a free-living setting [7]. In contrast, our study found high or moderately high intra- and inter-monitor reliability across a variety of ambulatory and non-ambulatory activities in both laboratory and in free-living settings. Thus, available evidence suggests that CPAM reliability may be stronger than CPAM validity; in other words, CPAMs may be more useful for tracking PA changes within an individual over time or comparing PA trends between individuals than assessing adherence to PA recommendations. This should be taken into consideration when determining the utility of CPAMs as tracking or intervention tools.

Weaker correlations were observed in isolated cases during the free-living data collection, likely as a result of large differences in predicted activity in a few participants (Figures 7–9). Interestingly, there is a noticeable discrepancy between these correlations and their respective MAD and MPD values. While these results seemingly contradict one another, the large differences which significantly impacted the correlations are not as influential in an analysis of the median absolute and percent differences because median values are not sensitive to outliers. The robustness of median values (compared to means) allows for better interpretation of differences between monitors and is supported by its use in previous research [15]. Nevertheless, the large differences and data removed per the 'cleaning' process are worth noting. It is the authors' impression that the artificial laboratory analysis and subsequently frequent uploading of CPAM data may have introduced some of these data (e.g., Figure 5). CPAMs are likely not intended to be updated in five-minute intervals over an extended period of time; thus, these errors could be attributable to application lag. It is worth noting, though, that these devices may have occasional errors while updating one or more variables. This may have contributed to some of the instances where the variables were actually lower at the end of the day than the beginning of the day (which is not physiologically possible). Additionally, some of the sporadic large differences seen in the free-living data (e.g., a difference of >6000 steps for the Fitbit Zip; differences of >375 kcals for the Fitbit One and Flex; differences of >10 active minutes for the Fitbit Zip and Jawbone Up24) may be attributable to occasional data loss during updating. Issues with updating the devices and/or associated applications are worth noting as they quantitatively lower reliability and may necessitate data screening or removal rules to be introduced.

A limitation of this study design was the relatively short duration (five minutes) of the laboratory activities which did not permit the analysis of active minutes (require at least 10-min bouts for Fitbit monitors). The abbreviated activity times may have also contributed to relatively frequent failure of CPAMs and/or their related applications to update properly resulting in bad data (removed from analysis). Additionally, sweeping was the only non-ambulatory, non-sedentary activity in the present study, which limits understanding of CPAM reliability during these types of activities. Mean wear time during the free-living portion of the study is also a limitation as it resulted in low data variability and limited options for statistical analyses for these data. To this, the limited wear time of these monitors in the free-living setting did not permit a statistical comparison between laboratory-based and free-living performance. Furthermore, a washout period was not utilized between laboratory and free-living segments of the study. This limitation introduces a source of variability such that participants may have modified their free-living behavior having completed the laboratory protocol earlier in the day. However, the laboratory activity protocol included a variety of activities not previously assessed in reliability studies (e.g., sweeping and reading) strengthens the present study. By including these activities of daily living, our results better reflect the performance of these monitors to across a variety of activities likely to be performed during a typical day. Second, this study included both laboratory and free-living aspects, which provides a more developed assessment of CPAM performance compared to studies without a free-living component.

In conclusion, these CPAMs provide reliable estimations of most PA variables in the laboratory; however, their reliability declines in a free-living setting. This may be attributable to small discrepancies between estimations being amplified as a result of increased wear time. Nonetheless, these findings suggest that certain CPAMs can provide reliable estimations of PA, especially steps taken, in a laboratory setting and possibly in free-living.

Acknowledgments: The authors would like to thank Reem Hindi and Gabriela Torres for assistance with data collection and subject recruitment.

Author Contributions: Joshua M. Bock, Leonard A. Kaminsky, Matthew P. Harber, and Alexander H. K. Montoye conceived and designed the study. Joshua M. Bock and Alexander H. K. Montoye were responsible for collecting data. Joshua M. Bock was responsible for data analysis. Joshua M. Bock and Alexander H. K. Montoye wrote the initial manuscript. Joshua M. Bock, Leonard A. Kaminsky, Matthew P. Harber, and Alexander H. K. Montoye revised and prepared the final manuscript.

Conflicts of Interest: The authors declare no conflict of interest.

References

1. Troiano, R.P.; Berrigan, D.; Dodd, K.W.; Masse, L.C.; Tilert, T.; McDowell, M. Physical activity in the United States measured by accelerometer. *Med. Sci. Sports Exerc.* **2008**, *40*, 181–188. [CrossRef] [PubMed]
2. Troiano, R.P.; McClain, J.J.; Brychta, R.J.; Chen, K.Y. Evolution of accelerometer methods for physical activity research. *Br. J. Sports Med.* **2014**, *48*, 1019–1023. [CrossRef] [PubMed]
3. Center for Disease Control Physical Activity Data and Statistics. Available online: http://www.cdc.gov/physicalactivity/data/ (accessed on 11 April 2017).
4. Wareham, N.J.; Rennie, K.L. The assessment of physical activity in individuals and populations: Why try to be more precise about how physical activity is assessed? *Inter. J. Obes.* **1998**, *22*, S30–S38.
5. Inside wearables Part 2. Available online: http://digitalintelligencetoday.com/wp-content/uploads/2015/11/2014-Inside-Wearables-Part-2-July-2014.pdf (accessed on 11 April 2017).
6. Worldwide Wearables Market Increases 67.2% Amid Seasonal Retrenchment, According to IDC. Available online: http://www.idc.com/getdoc.jsp?containerId=prUS41284516 (accessed on 11 April 2017).
7. Evenson, K.R.; Goto, M.M.; Furberg, R.D. Systematic review of the validity and reliability of consumer-wearable activity trackers. *Int. J. Behav. Nutr. Phys. Act.* **2015**, *12*, 159–181. [CrossRef] [PubMed]
8. Kooiman, T.J.M.; Dontje, M.L.; Sprenger, S.R.; Krijnen, W.P.; van der Schans, C.P.; de Groot, M. Reliability and validity of ten consumer activity monitors. *BMC Sport Sci. Med. Rehabil.* **2015**, *7*, 24–35. [CrossRef] [PubMed]
9. Diaz, K.M.; Krupka, D.J.; Chang, M.J.; Peacock, J.; Ma, Y.; Goldsmith, J.; Schwartz, J.E.; Davidson, K.W. Fitbit: An accurate and reliable device for wireless physical activity tracking. *Int. J. Cardiol.* **2015**, *185*, 138–140. [CrossRef] [PubMed]
10. Mammen, G.; Gardiner, S.; Senthinathan, A.; McClemont, L.; Stone, M.; Faulkner, G. Is this bit fit? Measuring the quality of the Fitbit step-counter. *Health Fit. J. Can.* **2012**, *5*, 30–39.
11. Takacs, J.; Pollock, C.L.; Guenther, J.R.; Bahar, M.; Napier, C.; Hunt, M.A. Validation of the Fitbit One activity monitor device during treadmill walking. *J. Sci. Med. Sport* **2014**, *17*, 496–500. [CrossRef] [PubMed]
12. Dontje, M.L.; de Groot, M.; Lengton, R.R.; van der Schans, C.P.; Krijnen, W.P. Measuring steps with the Fitbit activity tracker: An inter-device reliability study. *J. Med. Eng. Tech.* **2015**, *39*, 286–290. [CrossRef] [PubMed]
13. Chowdhury, E.A.; Western, M.J.; Nightingale, T.E.; Peacock, O.J.; Thompson, D. Assessment of laboratory and daily energy expenditure estimates from consumer multi-sensor physical activity monitors. *PLoS ONE* **2017**, *24*, e0171720. [CrossRef] [PubMed]

14. Murakami, H.; Kawakami, R.; Nakae, S.; Nakata, Y.; Ishikawa-Takata, K.; Tanaka, S.; Miyachi, M. Accuracy of wearable devices for estimating total energy expenditure: Comparison with metabolic chamber and doubly labeled water method. *JAMA Intern. Med.* **2016**, *176*, 702–703. [CrossRef] [PubMed]

15. Ferguson, T.; Rowlands, A.V.; Olds, T.; Maher, C. The validity of consumer-level, activity monitors in healthy adults worn in free-living conditions: A cross-sectional study. *Int. J. Behav. Nutr. Phys. Act.* **2015**, *12*, 42–51. [CrossRef] [PubMed]

16. Safrit, M.J.; Wood, T.M. *Introduction to Measurement in Physical Education and Exercise Science*; Mosby: St. Louis, MO, USA, 1995.

17. Swartz, A.M.; Strath, S.J.; Bassett, J.R.; O'Brien, W.L.; King, K.A.; Ainsworth, B.E. Estimation of energy expenditure using CSA accelerometers at the hip and wrist sites. *Med. Sci. Sports Exerc.* **2000**, *32*, S450–S456. [CrossRef] [PubMed]

18. Nelson, M.B.; Kaminsky, L.A.; Dickin, D.C.; Montoye, A.H.K. Validity of consumer-based physical activity monitors. *Med. Sci. Sport Exer.* **2016**, *48*, 1619–1628. [CrossRef] [PubMed]

19. Kamada, M.; Shiroma, E.J.; Harris, T.B.; Lee, I.M. Comparison of physical activity assessed housing hip- and wrist-worn accelerometers. *Gait Posture* **2016**, *44*, 23–28. [CrossRef] [PubMed]

20. Fairclough, S.J.; Noonan, R.; Rowlands, A.V.; van Hees, V.; Knowles, Z.; Boddy, L.M. Wear compliance and activity in children wearing wrist- and hip-mounted accelerometers. *Med. Sci. Sports Exerc.* **2016**, *48*, 245–253. [CrossRef] [PubMed]

Review

Cuff-Less and Continuous Blood Pressure Monitoring: A Methodological Review

Manuja Sharma [1], Karinne Barbosa [1], Victor Ho [1], Devon Griggs [1], Tadesse Ghirmai [1], Sandeep K. Krishnan [2], Tzung K. Hsiai [3], Jung-Chih Chiao [4] and Hung Cao [1,*]

[1] Division of Engineering and Mathematics, School of Science, Technology, Engineering and Mathematics, University of Washington, Bothell, WA 98011, USA; manuja21@gmail.com (M.S.); karinnejcbarbosa@gmail.com (K.B.); vhau422@gmail.com (V.H.); devongriggs@outlook.com (D.G.); tadg@uw.edu (T.G.)

[2] Division of Interventional Cardiology, Department of Medicine, School of Medicine, University of Washington, Seattle, WA 98195, USA; sdizzle@cardiology.washington.edu

[3] Division of Cardiology, David Geffen School of Medicine, University of California, Los Angeles, CA 90095, USA; THsiai@mednet.ucla.edu

[4] Department of Electrical Engineering, University of Texas Arlington, Arlington, TX 76019, USA; jcchiao@uta.edu

* Correspondence: hungcao@uw.edu; Tel.: +1-425-352-5194

Received: 5 March 2017; Accepted: 5 May 2017; Published: 9 May 2017

Abstract: Blood pressure (BP) is one of the most important monitoring parameters in clinical medicine. For years, the cuff-based sphygmomanometer and the arterial invasive line have been the gold standards for care professionals to assess BP. During the past few decades, the wide spread of the oscillometry-based BP arm or wrist cuffs have made home-based BP assessment more convenient and accessible. However, the discontinuous nature, the inability to interface with mobile applications, the relative inaccuracy with movement, and the need for calibration have rendered those BP oscillometry devices inadequate for next-generation healthcare infrastructure where integration and continuous data acquisition and communication are required. Recently, the indirect approach to obtain BP values has been intensively investigated, where BP is mathematically derived through the "Time Delay" in propagation of pressure waves in the vascular system. This holds promise for the realization of cuffless and continuous BP monitoring systems, for both patients and healthy populations in both inpatient and outpatient settings. This review highlights recent efforts in developing these next-generation blood pressure monitoring devices and compares various mathematical models. The unmet challenges and further developments that are crucial to develop "Time Delay"-based BP devices are also discussed.

Keywords: blood pressure; pulse transit time; pulse arrival time; electrocardiogram (ECG); photoplethysmography (PPG)

---

1. Introduction

Cardiovascular disease (CVD) plagues our aging society as the leading cause of morbidity and mortality in developed countries [1,2]. High blood pressure (BP) or hypertension (HTN) is a common condition leading to CVD. HTN is determined by increased pressure in the arteries that can lead to stress on the heart, also known as hypertensive heart disease. About 67 million American adults (31%) are affected by HTN, while only 47% of patients maintain normal BP control [3]. Further, HTN has also been found associated with other health issues in various groups of populations, such as the elderly and pregnant women, to name a few [4,5].

Conventionally, non-invasive BP has been measured using a sphygmomanometer based on the design proposed by Samuel Siegfried Karl Ritter von Basch in 1881 [6]. Riva Rocci further improved the design by developing a branchial cuff sphygmomanometer in 1896 [7]. The detection of Kortokoff sound (K-sound) in 1905 enabled complete non-invasive BP measurement [8]. The pressure indicated by the manometer at the first K-sound is noted as the systolic BP (SBP) and the silent fifth sound indicates the diastolic BP (DBP) (Figure 1a) [8–10]. In the past few decades, oscillometry-based BP tools have become popular, providing ease of operation. They do not require a caregiver or experienced personnel to operate and hence can be used to monitor BP in the home setting. These devices have a cuff wrapping around the arm or leg to detect the oscillations during cuff-deflation using a built-in pressure sensor. Mean arterial pressure (MAP) is estimated using the amplitude variations of the recorded oscillations which are used to algorithmically obtain SBP and DBP [11,12] (Figure 1b). However, a recent study by Leung et al. indicated over three in ten home BP monitoring cuffs were inaccurate [13]. Further, cuff-based devices are cumbersome and cannot perform continuous measurements. Thus it is difficult to be integrated with wearable technologies, which continue to gain popularity in commercial sectors and clinical practice. Ambulatory blood pressure monitoring (ABPM) has been used to diagnose HTN in the outpatient setting. Although, ABPM is superior to the isolated, sporadic monitoring of patient's BP generally affected by "white coat" HTN (or "white coat syndrome", referring to a phenomenon in which people only exhibit HTN in a clinical setting), the current ABPM in use is a bulky device that is not portable or practical for daily or long-term uses [14].

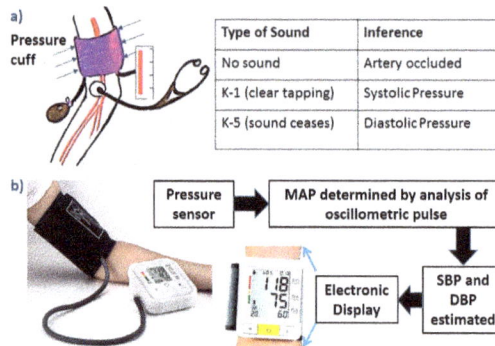

Figure 1. Conventional blood pressure (BP) measurement. (**a**) Sphygmomanometer; and (**b**) Oscillometry-based BP measurement.

Finapres (*Finapres Medical Systems*, the Netherlands), a device that measures finger arterial pressure using a finger cuff and infrared plethysmograph, has been gaining in popularity [15,16]. Though having a smaller cuff, the processing device still makes it inadequate to provide continuous data for daily use. Additionally, it is also motion-sensitive and cannot be reliably used to measure BP during normal activities.

In the past few years, several research groups have developed cuff-less BP monitoring wearable devices, holding promise to allow patients to continuously monitor BP without interruption to their daily activities [17–21]. The underlying principle of these devices is based on the relation of the time it takes for a volume of blood (in the form of a pulse) to travel from the heart to a peripheral organ, which could be in the form of pulse transit time (PTT) or pulse arrival time (PAT) [22–24]. Algorithms and mathematical models have been proposed and developed to optimize the regression process and calibration of the traveling/delay time ("Time Delay") and BP [25–27]. The "Time Delay" is usually obtained using a cardiac electrical signal, i.e., electrocardiogram (ECG), recording device and a pulse oximeter at a peripheral organ, i.e., photoplethysmography (PPG) [28,29]. However, existing systems are not continuous as most of the ECG acquisition approaches require a cross-body configuration, asking the user to touch an open electrode on the wearable device; hence continuous measurements of BP have not been achieved [17,30]. Though these devices overcome issues of other non-invasive tools, several critical issues still remained. First, some devices anchor to the body, which some users may find irritating [31]. Second, this approach requires frequent calibration to map the "Time Delay" and BP to maintain accuracy and all existing fitting models appear to be dependent of objects and temporal trials, as well as motional activities [32,33]. Third, time synchronization between different bio-signals to obtain "Time Delay" is crucial as the system's input parameter is in milliseconds. Fourth, the dependence of BP on other factors, such as vasomotor tones, neural control and heart rate, requires additional parameters to be included along with PTT/PAT in the mathematical model to adequately estimate BP [34,35]. Lastly, the accuracy measured via the regression coefficient (R^2) is low, with significant value variations even in the same subject at the same activity level [36,37], suggesting a more-sophisticated mathematical model may be a solution to enable BP monitoring devices with higher reliability and precision, which can later be accepted as practical medical-grade tools.

This methodological review provides an overview of the physical relationship between blood pressure and the "Time Delay" of cardiac signals in a human's cardiovascular system, as well as the existing mathematical models to derive BP from "Time Delay" measurements. Thorough comparisons among methods are achieved using our recording data [38] and those obtained from the Physionet's Multi-parameter Intelligent Monitoring in Intensive Care (MIMIC) II (Version 3, accessed in December 2016) online waveform database [39]. Collectively, challenges and important issues of wearable, home-based, cuffless and continuous BP monitoring are discussed in detail.

2. Cardiovascular System and the Electrical-Mechanical Coupling of the Heart

2.1. Electro-Mechanical Cardiac Signals

The heart's electrical system, also known as the cardiac conduction system, consists of three main components: sinoatrial (SA) node, atrioventricular (AV) node and His-Purkinje system. It is usually recorded as the ECG signal [40] (Figure 2a). The SA node, located in the upper portion of the right atrium, is the heart's intrinsic pacemaker that initiates electrical signal, indicated as the P-wave of the patient's ECG. Generated electrical signals result in atrial contraction and help push blood through the atrioventricular valves into both ventricles. The electrical impulse then activates the AV node, a relay station, situated above the ventricles. It facilitates right and left atriums to empty their blood contents into the two ventricles (corresponding to PR intervals of ECG). Once released, the electrical signal moves along the electrical highway (the "bundle of His"—transmits impulses from the atrioventricular node to the ventricles of the heart) which later divides into Purkinje fibers connected to cells in the walls of the left and right ventricles. This causes the electrically stimulated ventricles to contract and pump oxygenated blood into the arteries. This entire phase represents the QRS complex of an ECG. Later stages of the ECG signal represent the repolarization phase of the ventricles (T waves) [41,42].

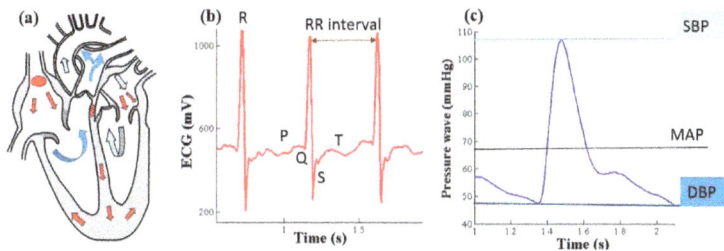

Figure 2. (**a**) Electro-mechanical signal generated in human heart; (**b**) electrocardiogram (ECG) signal; and (**c**) pressure wave.

Electrical-mechanical coupling of the heart results in blood ejection into the arterial tree, affecting the blood velocity and generating a systemic pressure wave traveling from the central to peripheral arteries. The pressure wave causes dilation of the arterial walls on its path and moves faster than the blood flow [43,44]. It varies periodically between two extreme points, maximum and minimum, referring to SBP and DBP, as the pressure in the artery due to ventricular contraction and the pressure in the artery during each beat, respectively. The mean value of the pressure wave, termed MAP (Figure 2c), is estimated as:

$$MAP = DBP + \frac{1}{3}(SBP - DBP) \tag{1}$$

This pressure waveform can be directly obtained using a pulse sensor on the peripheral arteries or indirectly measured through a pulse oximeter, namely a PPG sensor. Since the pressure wave causes the blood volume to change at the peripheral site, it can be detected by measuring the variation of the oxygen content of the blood caused by influx of oxygenated blood on the arrival of the pressure wave, indicated as the first peak on the PPG waveform [45]. Many other vital parameters can be estimated using PPG as discussed in section C.

2.2. Relationship between BP, Pressure Wave Velocity (PWV) and Time Period

The central arteries push blood to narrow distal arteries by expanding during systole and contracting during diastole [46]. This expansion and contraction results in changes of the elastic modulus (E) of the vessels and is related to the fluid pressure P as below:

$$E = E_0 e^{\alpha P} \tag{2}$$

In Equation (2), α is a vessel parameter (Euler number) and E_0 is the Young's modulus for zero arterial pressure. These two are subject-specific parameters [47–49]. Equation (2) estimates the central arterial pressure if α and E_0 are updated by accounting for the age and health impacts on the elasticity due to the change in the wall composition. Arterial walls are composed of endothelium, elastin, collagen, and smooth muscle (SM) cells in varying quantities at central and peripheral sites [50,51]. Different compositions as well as gradual replacement of elastin with collagen changes the elasticity of these arteries, resulting in changes in central and peripheral BP [52,53]. A detailed analysis of elastic and viscous properties of the arterial tree can be found in the review [54] which describes how the central arterial elasticity is determined by the BP and also how the peripheral elasticity is affected by both BP and SM contraction [55]. Hence, the peripheral elasticity cannot be accurately predicted by Equation (2).

The elasticity of arteries determines the propagation speed, the pressure wave velocity (PWV); a relationship can be obtained between them using arterial wave propagation models. Assuming the artery to be an elastic tube with a thickness h, diameter d and blood density ρ, we have the Moens-Kortweg equation as follows [56]:

$$PWV = \sqrt{\frac{hE}{\rho d}} \tag{3}$$

Combining Equations (2) and (3), we obtain the Bramwell-Hills and Moens-Kortweg's equation, representing the relationship between P and PWV and hence the "Time Delay" for an artery with a length of L [44]:

$$PWV = \frac{L}{\text{Time Delay}} = \sqrt{\frac{hE_0 e^{\alpha P}}{\rho d}} \tag{4}$$

This equation indicates that the rise in pressure, with other parameters constant, will result in an increase in PWV and inversely affects the "Time Delay".

2.3. Determination of the "Time Delay"

2.3.1. Pulse Transit Time (PTT)

PTT refers to the time taken by a pressure wave to travel between two arterial sites and is inversely related to BP (Figure 3a). PTT can be measured using different techniques like Ultrasound Doppler and arterial tonometry [57–59]. The latter can be obtained by observing two distant PPG waves (Figure 3c). Ears, toes and fingers are common sites used for measurement [60]. PTT_f measured from the foot of one PPG to that of another has been demonstrated to have a strong correlation to invasive DBP [61], but a study on 44 normotensive male subjects concluded otherwise [62]. The peaks of PPG that theoretically represent SBP have been found to be unreliable indicators of SBP. These peaks are distorted by reflection of pressure waves from the terminal arteries. Chen et al. proposed a novel method based on experimental data to use the mean of PTT_{fp} (time delay between falling edge of central PPG and peak of peripheral PPG) and PTT_{rp} (time delay between rising edge of reflecting central PPG and peak of peripheral PPG) to obtain PTT for SBP (Figure 3c) [63]. Other studies have indicated that posture, ambient temperature, and relaxation affect PPG, raising a question on the development of an accurate PTT device [64].

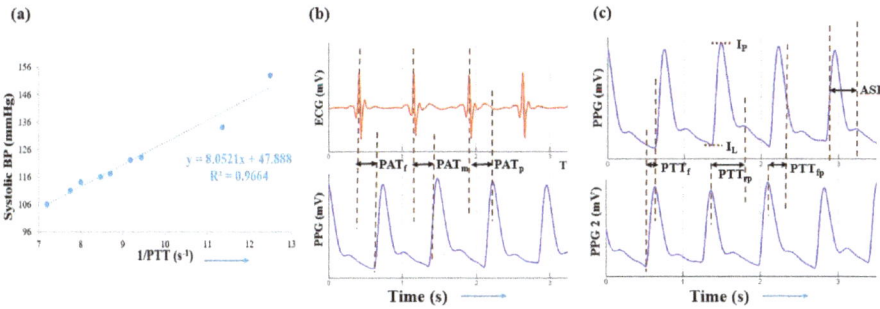

Figure 3. (**a**) Inverse relationship between pulse transit time (PTT) and systolic blood pressure (SBP) from our data; (**b**) Pulse arrival time (PAT) using different characteristic points of the photoplethysmography (PPG) waveform; (**c**) Different types of PTT and other PPG parameters.

2.3.2. Pulse Arrival Time (PAT)

Another popular and convenient method to measure the "Time Delay" is based on the time difference between the R-peak of ECG and a characteristic point of PPG peak (Figure 3b). Different time stamps on the PPG waveform, such as foot [65–67], peak [67–69] and mid-point of the rising edge [29,70,71], have been considered to estimate the "Time Delay" (Figure 3b). Though some studies reported this delay as PTT, it is more accurately known as Pulse Arrival Time (PAT) as in addition to the PTT of the pressure wave, it includes the Pre-ejection Period (PEP) delay. PEP is the time needed to convert the electrical signal into a mechanical pumping force and isovolumetric contraction to open the aortic valves [43,72]:

$$PAT = PTT + PEP \tag{5}$$

PEP is a delay that changes with stress, physical activity, age and emotion [48]. A study has attempted to estimate PEP as a percentage of the RR interval as, with a low heart rate, PEP becomes more significant [73]. They approximated PEP as 7% of the RR interval and concluded that it should be subtracted out to obtain PTT. The impact of PEP on the overall PTT decreases with distance from the heart. Thus for short PTTs, especially those extracted from ear PPQ, it is needed to accommodate for this electro-mechanical delay, or PEP. Nevertheless, the effect of including PEP in BP estimation is still under investigation. Some have reported the relationship of PEP with PAT [74–76], while others find it a weak surrogate [77,78]. There were studies indicating that all SBP, DBP and MAP are less correlated to PAT as compared to PTT [79], while others stating PAT is a better indicator of SBP [80,81] as it is dependent on both ventricular contraction and vascular function.

3. Mathematical Models

Mathematical relationships between BP and the "Time Delay" or PTT/PAT reported in the literature are derivatives of the physical model previously discussed and summarized in Table 1. We will discuss several models below.

Table 1. Summary of mathematical models to calculate BP from PTT/PAT.

Algorithm	Time Delay	Calibration Technique	Results SBP	Results DBP	Results MAP	Refs
$a\ln(\text{Time Delay}) + b$	PTT_f		-0.22 ± 0.46 [2]	-	-	[82]
	PAT_f	Subject specific	-0.85 ± 0.09 [2]			
	PTT_f			0.9 [3]		[73]
$\dfrac{A}{PAT^2} + B$	PAT_p	A: Dependent on Height B: Subject specific	0.0790 ± 11.32 [1]	-	-	[73]
	PAT_f	Subject specific	$0.99\text{–}0.90$ [2]	-	-	[65]
	PAT_p	Subject specific	-0.92 [2]	-0.38 [2]	-	[83]
			RT: -0.87 [2,*]	RT: -0.30 [2,*]		
	PAT_m	Subject specific	0.701 [3]	0.401 [3]	-	[84]
	PAT_p	Subject specific	$0.95\text{–}0.87$ [3]	$0.01\text{–}0.73$ [3]	-	[85]
$aPAT + b$	PAT_p	Subject specific	-0.71 [2]	-0.69 [2]	-	[25]
	PAT_m		-0.32 [2]	-0.22 [2]		
	PAT_f		-0.09 [2]	-0.02 [2]		
	PAT_p	Subject specific			0.32 [2]	[28]
	PAT_m				0.27 [2]	
	PAT_f				0.45 [2]	
$\dfrac{a}{PAT} + b$	PAT_p	Subject specific	0.95 [3]	0.26 [3]		[69]
	PAT_m		0.89 [3]	0.78 [3]		[86]
$\dfrac{a}{PAT} + b + c*VPAT + d*(PATV - PATV_0)$	PAT_p		0.96 [3]	0.70 [3]		[69]
$a + \left(\dfrac{b}{PAT-c}\right)^2$	PAT_p	Subject specific	0.97 ± 0.87 [3]	0.54 ± 0.05 [3]	-	[85]
$\dfrac{(PWV-a)}{b}$	PAT_f	Subject specific	0.93 [2]	-	0.83 [2]	[87]
	PAT_f		0.94 [2]		0.86 [2]	
$aPAT + bHR + c$	PAT_m	Subject specific, Maximum Likelihood	0.978 [3]	0.974 [3]	-	[88]
	PAT_m	Subject specific, Adaptive Kalman filter	0.976 [3]	0.989 [3]	-	
$a + bPAT + cHR + dTDB$	PAT_m	Subject specific	0.85 [3]	0.74 [3]	-	[84]

Table 1. *Cont.*

Algorithm	Time Delay	Calibration Technique	Results			Refs
			SBP	DBP	MAP	
$BP_{ij} = b_{ij}e^{-\left(\frac{K_{ij}}{PWV_{ij}}\right)}$ $i = 1,2,\dots,m(\text{age})$ and $j = \text{Male}/\text{Female}$	$\dfrac{PTT_f}{PTT_{fp} + PTT_{fp}}$	b_{ij} and K_{ij} are calculated for a demographically similar group.	2.16 ± 6.23 [1]	-1.49 ± 6.51 [1]	-	[61,63]
$DBP = \frac{SBP_0}{3} + \frac{2}{3}\frac{DBP_0}{3} + a\ln\left(\frac{PAT_{90}}{PAT_w}\right) - \frac{(SBP_0-DBP_0)}{3}\left(\frac{PAT_{90}}{PAT_w}\right)^2$ $SBP = DBP_0 + (SBP_0 - DBP_0)\left(\frac{PAT_{90}}{PAT_w}\right)^2$	PAT_m	Subject specific	0.6 ± 9.8 [1]	0.9 ± 5.6 [1]	-	[30]
$BP = a*PWV*e^{b*PMV} + cPWV^d - (BP_{PTT,cal} - BP_{PTT})$	PAT_m	Universal	0.83 [2]	-	-	[89]
$DBP = DBP_0\frac{PTR_0}{PTR}$ $SBP = DBP_0\frac{PTR_0}{PTR} + PP_0\left(\frac{PAT_0}{PAT}\right)^2$	PAT_m	Subject specific	0.91 [2]	0.88 [2]	0.89 [2]	[90]

[1] Mean ± SD; [2] r: Correlation Coefficient; [3] R^2: Coefficient of Determination; * RT: Repeatability Test.

3.1. Logarithmic Model

The Bramwell-Hills and Moens-Kortweg's equation gives a logarithmic relationship between BP and the "Time Delay". Assuming the density of blood (ρ), the diameter of artery (d), the thickness of the artery (h), the distance at which the "Time Delay" is obtained (L), and the elasticity (E_o) are constant for a subject, we can have relationship of BP and the "Time Delay" represented as:

$$BP = a \, \ln(\text{Time Delay}) + b \qquad (6)$$

Here, a and b are subject-specific constants and they can be obtained through a regression analysis between the reference BP and the corresponding "Time Delay" [82]. Proença et al. estimated SBP with this mathematical model Equation (6) using both PTT and PAT [82]. They determined PTT from two PPG sensors placed at the earlobe and at a finger, and PAT with PEP adjustment using the impedance cardiogram. However, they found inconsistent results with both of them. Poon et al. established a relationship between MBP and the "Time Delay" using Equation (6) and obtained SBP and DBP using Equation (1) and a factor that accounts for the change in elasticity due to pressure wave variations [30]. Their results agreed with the AAMI (American Association for the Advancement of Medical Instrumentation) standard of a BP device with the mean difference of less than 5 mmHg and standard deviation within 8 mmHg [91]. Hence, this method has become popular to indirectly obtain BP via the "Time Delay".

The logarithmic model Equation (6) approaches negative infinity as "Time Delay" tends to zero, making it difficult to use this relationship to represent small BP [61].

3.2. Proportional (Linear) Model

Assuming there is a negligible change in the arterial thickness and diameter with pressure variations, BP and the "Time Delay" can be linearly related by differentiating the Moens-Kortweg's Equation (3) with respect to time Equation (6) [65]. Chen et al. obtained a high correlation factor between the measured SBP and the calculated SBP using PAT_f and thus they established a calibration model that varies according to fluctuation in PAT.

$$BP = a(\text{Time Delay}) + b \qquad (7)$$

Using Equation (7), a study attempted to estimate SBP and DBP for 14 normotensive subjects and then carried out a repeatability test after six months to verify whether the model still holds or not [83]. In their study, PAT_p was obtained for each subject before, during and after exercise. The repeatability test showed that although the range of a and b were similar to the values obtained previously, errors on calculating BP using the *"six-month old calibrated algorithm"* were significant. The authors also reported that there was less correlation between DBP and PAT which could be due to the fact that PAT_p, instead of PAT_f, was used to estimate DBP. Choi et al. used the same algorithm and investigated the use of different characteristic points of PPG (Figure 3b) and calibration intervals to achieve BP [25]. In their work, they concluded that PAT_p measurements and one hour calibration intervals provided better estimates of BP within error limits. Further, it was claimed that the algorithm was adopted not only for its better performance, but also for its robustness against motion artifacts that exist in non-invasive waveforms.

Several other studies have integrated the linear BP algorithm (7) with other influencing factors, such as heart rate (HR) and arterial stiffness index (ASI) (Figure 3c), that would affect BP [48,66,84]. The effect of variance in HR has shown both positive and negative impacts on BP in clinical data. In normal conditions, it has a positive relation but under baroreflex activity (the mechanism to regulate acute BP changes via controlling heart rate), HR is negatively correlated to BP [84,87,92]. The other factor, arterial stiffness, has been assumed constant in algorithms based on those physical models (3) and (4). However, it influences the calibration frequency and can be estimated using ASI. The correlation factor using HR and the linear model (7) has been found to be around 0.79 for SBP and 0.814 for DBP [84], confirming its significance. It was also found that estimations based on the "maximum likelihood" and adaptive Kalman filter can reduce the number of calibration measurements required to estimate algorithm constants [88]. This model can be estimated by linear regression and offers ease in recalibration.

3.3. Inverse Square Model

Assuming arteries are rigid pipes, the work done by the travelling pressure wave can be expressed as a sum of its potential and kinetic energy where kinetic energy is dependent on PWV. The work done is equal to the change in BP with a fixed cross-sectional area and thus [73]:

$$BP = \frac{A}{\text{Time Delay}^2} + B \tag{8}$$

where $A = \left(0.6 \times \frac{\text{height}}{\text{distance factor}}\right)^2 \frac{\rho}{1.4}$ and ρ is the average blood density.

Here, B is estimated for individuals by using the "Time Delay" and cuff-based BP measurements. Fung et al. used PAT to estimate BP, assuming that the measurement obtained from peripheral sites like toes and fingers have insignificant delay due to PEP. However, for PPG at the ear, PAT should have been adjusted for PEP as the ear is closer to the heart, thus possessing a pressure wave similar to that in the central artery. For the estimation of A from the subject's height, an additional distance factor related to the locations of the PPG sensors was included. The distance factor was assumed to be 1, 0.5, and 1.6 for fingers, ears, and toes, respectively, in the study. The algorithm correlated to BP measured by cuff with a mean difference of −0.0790 mmHg and 11.32 mmHg standard deviation. Furthermore, it was able to track both hypotension and hypertension. Wibmer et al. [85] modified the above relationship to account for the asymptotic behavior of BP as following:

$$BP = a + \left(\frac{b}{\text{Time Delay} - c}\right)^2 \tag{9}$$

In their study, PAT was obtained using PPG and a single-lead ECG signal and it had a high correlation factor with SBP while the correlation with DBP was similar as previously reported. Thus, the approach (9) adds reasonable asymptotic behavior which most models fail to achieve.

3.4. Inverse Model

The model represented by (9) also indicates the inverse relationship between BP and PTT and thus was used to obtain the subject-specific mathematical equation [54,93], where BP is calculated as follows

$$BP = \frac{a}{\text{Time Delay}} + b \tag{10}$$

Additional parameters accounting for neural impact on BP were included to the inverse relationship given by (10). It has been reported that variability in both PTT and BP signals was coherent indicating that the neural system affects them simultaneously. When compensation for variability due to neural control was integrated with (10), the model gave results with higher accuracy for both DBP and SBP [69]. The study also incorporated the hydrostatic effects in the algorithm by measuring data for calibration in sitting and standing position. Model (10), if represented in terms of PWV, gives a direct relationship between PWV to BP [87]. In this case, the parameter L which is the distance between the sensor point and heart, was obtained using a subject-specific tape measure between the *fossa jugularis* and the sensor instead of relying on various ratios to subject's height. Marcinkevics et al. considered two PWV estimation methods, using PAT_f and PTT_f, and obtained similar results [87]. Most experimental data suggest the inverse relationship between BP and PTT/PAT which is achieved by this model.

3.5. Comparison between Models

It was reported that those algorithms previously discussed yielded R^2 values ranging from 0.02 to 0.97 as summarized in Table 1. In order to apply a mathematical model, one needs to vary the BP over a considerable range to obtain the curve that can relate PTT/PAT closely to BP. Models under different conditions, generally exercising and medication, have been considered as summarized in Table 2. The same algorithm may result in different regression coefficients when using different BP perturbation techniques and calibration intervals, thus making comparisons difficult [25,84]. In order to give a better insight on this issue, in this work we elucidated four algorithms (6)–(8), and (10), using the Physionet online database as a source for the ECG, PPG and Arterial Blood Pressure (ABP). The database contains time-stamped nurse-verified physiological readings of patients in the intensive care unit at Boston's Beth Israel Deaconess Medical Center (BIDMC), beginning in 2001 and spanning seven years [94]. Individual demographic details, though vital in analyses, are unavailable as it can result in infringement of patient's privacy [95]. Two-leads ECG, finger PPG, and invasive ABP from one of the radial arteries were recorded at 125 Hz for varied lengths (weeks or more) using a bedside monitor. In this review, data which had at least five-minute long continuous recordings were considered for investigation. The ABP values listed were used for the calculation of SBP, DBP, and MAP. The four algorithms were evaluated using PAT_f, PAT_m, and PAT_p for five adult patients (age > 15) by using five different PATs and the average of five beat-to-beat ABP values as a reference. The unavailability of a second PPG signal restricted our evaluation to PAT. The Band-Altman plot for the expected measurement (BP values calculated from invasive ABP waveform) versus the calculated BP as well as R^2 were used to analyze the four algorithms (Figure 4) [96–98]. The *x*-axis of the plots shows the average estimate of the algorithm that performed best and the *y*-axis represents the difference between the expected and measured values. Results indicated that the inclusion of HR in the BP algorithm for all three BP values (SBP, DBP, and MAP) gave higher regression coefficients between the measured and expected values with mean error and standard deviation (SD) within AAMI standards (Figure 4), and the use of PAT_m and PAT_p gave similar or better results than PAT_f. Using this algorithm, 92.3% MAP data is within the interval of 2·SD and 85% of DBP estimates. In case of SBP, though using PAT_p gave better R^2, that of PAT_m delivered a more-agreeable Bland-Altman plot with 96.1% data within the 2·SD interval (Figure 4). For both DBP and MAP, the mean of difference between the invasive BP and the one estimated using PAT and HR is small, indicating insignificant bias between the two methods. Average error and standard difference between the PAT based and the reference value is within AAMI standards for both MAP (0 ± 2.12) and DBP (0 ± 2.13) measurements. The SBP measurement has a slightly higher standard difference (1.3 ± 7.02), but it is still within the AAMI standard (Table 2). The larger disagreements between reference and measured SBP and DBP values for some of the data points is due to inclusion of data of hypertensive patients, as can be seen in Figure 4a,c. Out of the five patients, two reported BP higher than 140/80 mmHg, resulting in higher disagreements in their case, which agrees with the study performed by Gesche and colleagues [89].

Table 2. Various BP perturbation techniques.

Procedure	Description	Refs
Physical Exercise	Graded Bicycle Test, Running, Sit-ups	[87,99]
Posture	Sitting, standing and lying supine	[100]
Valsalva Maneuver	Breathing against closed nose/mouth for 30 s	[101]
Cold Pressure	Placement of ice wrapped in wet cloth on subject's forehead for 2 min. Hand in 4 °C water for 1 min	[102,103]
Mental Arithmetic	Counting backwards from 500 in intervals of 7, Continuous addition of 3 digit numbers for 2 min	[104]
Relaxation	Slow breathing/Meditative music	[105,106]
Amyl Nitrate	Inhalation of vasodilator	[107]
Anesthesia	Dental anesthesia	[84,108]
Isometric Exercise	Raising legs/arms against pressure	[109]
Sustained Handgrip	Clenching one's fist forcefully	[110,111]

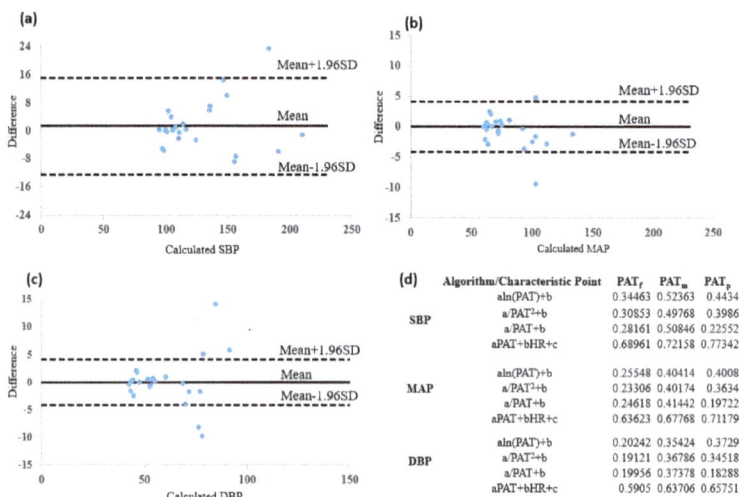

Figure 4. Bland-Altman plot between reference and measured (**a**) SBP using aPATm $+ b$HR $+ c$; (**b**) Mean arterial pressure (MAP) using aPATp $+ b$HR $+ c$; (**c**) Diastolic blood pressure (DBP) using aPATp $+ b$HR $+ c$; and (**d**) R^2 obtained for different algorithms with significance level 0.05.

Another study [112] provided a comparison between algorithms by Chen et al. and Poon et al. and they found that Chen's algorithm gave results within the standards for only a 4 min calibration interval and was unreliable in tracking large changes in BP, whereas Poon's algorithm required shorter calibration intervals to maintain a favorable accuracy. A recent study claimed that the inaccuracy of existing algorithms using PAT in tracking Low-Frequency (LF) variations in BP was one of the main reasons for inaccurate results [90,113]. In order to improve the accuracy, they introduced a factor termed "*photoplethysmogram intensity ratio*" (PIR), which could be determined by the ratio of the peak (I_p) and the foot (I_f) PPG values (Figure 3c). PIR was found mainly dependent on the arterial diameter and correlated with DBP [90,113]. Recently, machine learning-based techniques have been investigated [114] and promising results have been reported. In their work, various machine learning-based techniques were compared and they found that non-linear algorithms, like kernel machines or ensemble-learning methods as well as the AdaBoost model, gave better performances than linear approximations. Other techniques such as linear regression and decision tree were found not appropriate in predicting BP [114].

4. Discussion

Blood pressure is a complex parameter that has both physiological and neurological influences, and thus those need to be included to obtain a robust model. Heart rate (HR) represents the cardiac cycle and determines the heart's preload and the cardiac output (CO), which positively impact BP as the pressure on the arterial walls. HR is proportional to the volume of blood ejected [115,116]. HR and BP are also regulated by the autonomic nervous system which has been found to be inversely related, depending on the baroreflex activity [117]. HR is calculated from the RR interval in ECG signals and has been incorporated in several algorithms to estimate BP [84], demonstrating some improvement in accuracy. To consider the effect of the sympathetic system on BP, variability in PAT has been employed in the inverse model giving R^2 of 0.96 when compared to those of cuff-based SBP tools [69]. The arterial stiffness affecting cushioning of arteries has been found to have implications on systolic hypertension [53,118]. Structural changes, such as those related to vascular aging and atherosclerosis, along with functional changes like increased BP or higher sympathetic activity, can impact the arterial stiffness [115]. Baek et al. described a method of estimating arterial stiffness index (ASI) as the time delay between the mid-point of the rising edge (incident) and characteristic point on the "dicrotic notch" (reflecting wave) of the PPG (Figure 3c) [84,115]. Shaltis et al. estimated the impact of hydrostatic pressure by varying the position of subject and measuring changes in PTT and BP, and through experimental data they concluded that gravity has a significant impact on BP [119]. These suggest that one may need to take more related parameters under consideration in order to establish an accurate and robust mathematical algorithm/model/relation to derive BP from the "Time Delay".

There are several other important issues that make time delay-based BP measurements challenging. To find the best fitting curve for PTT/PAT-based BP estimation, it is essential to vary BP over a wide range in order to have more points to construct the calibration curve. Significant variations in SBP can be achieved using some of the methods listed in Table 3, but it is difficult to vary DBP and MBP, restricting the development of accurate algorithms for these. Another issue arises due to the assumption of a tubular arterial system similar to central arteries which fails for the tapering peripheral branches. Peripheral branches have amplified pressure peaks due to wave reflection from arterial terminations, which need to be adjusted to estimate central artery pressure [120,121].

Table 3. Mean error and standard deviation (in mmHg) between measured and reference BP data.

	Algorithm/Characteristic Point	PAT_f	PAT_m	PAT_p
SBP	$a \ln(PAT) + b$	0.1 ± 11	8.9	0.3 ± 11
	$a/PAT^2 + b$	0.1 ± 12	0.1 ± 12	0.1 ± 12
	$a/PAT + b$	0.1 ± 13	0.1 ± 13	0.1 ± 13
	$aPAT + bHR + c$	0.1 ± 14	0.1 ± 14	0.1 ± 14
MAP	$a \ln(PAT) + b$	0.1 ± 15	0.1 ± 15	0.1 ± 15
	$a/PAT^2 + b$	0.1 ± 16	0.1 ± 16	0.1 ± 16
	$a/PAT + b$	0.1 ± 17	0.1 ± 17	0.1 ± 17
	$aPAT + bHR + c$	0.1 ± 18	0.1 ± 18	0.1 ± 18
DBP	$a \ln(PAT) + b$	0.1 ± 19	0.1 ± 19	0.1 ± 19
	$a/PAT^2 + b$	0.1 ± 20	0.1 ± 20	0.1 ± 20
	$a/PAT + b$	0.1 ± 21	0.1 ± 21	0.1 ± 21
	$aPAT + bHR + c$	0.1 ± 22	0.1 ± 22	0.1 ± 22

Training algorithms require simultaneous recording of ECG/PPG and a gold standard tool for BP measurements. In many studies, this was carried out on different arms; however, BP is known to differ on both arms [122,123] and hence the correlation between this BP and PTT/PAT had inherent errors. The inter-arm BP difference is usually not taken in account while doing the comparison, which can induce errors [124]. The PPG signal which has been widely used to estimate PTT and PAT is susceptible to motion artifacts and thus requires careful placement of sensors and signal processing techniques, such as the periodic moving average filter [125–127]. The PPG sensors need to be fixed accurately on the arterial position for BP derivation, as modifying the position would change the distance of propagation, and thus affect PWV and ultimately BP [128]. The contact force of PPG sensors at the site of placement has a considerable effect on the BP measurement and needs to be adjusted during calibration [129]. A study demonstrated that at positive transmural pressures, PTT increased with the applied contact force, reaching the maximum at the zero transmural pressure and remaining at a constant level at negative transmural pressures [130]. Further, the detection of various characteristic points on a PPG signal requires effective signal processing algorithms as it is susceptible to noise and motion [131–134]. Though several studies have attempted to calibrate BP for dynamic cases, most of them required subjects to stop exercising to estimate BP and therefore failed to provide real-time data [26]. Movements of arms and legs would affect the PPG signal, and it is thus advised to include hydrostatic effects on the PTT-BP mathematical model. Liu et al. demonstrated that the same BP can have different PAT, depending on whether the subject is at rest or exercising, possibly due to the differences in PEP [35]. Accelerometers can be used to determine different body positions (sleeping/walking/running, etc) and provide additional information in selecting appropriate algorithms to estimate BP [17,130,135]. Studies have shown that longer PTTs give better estimations than shorter ones, and this requires sensors to be placed far away from each other; consequently, compact devices would be difficult to achieve [136]. There is also a need to include heart rate variability (HRV), as it has been shown that HR can have both negative and positive effects on BP, and hence the same algorithm will fail to accurately predict BP if the variability is ignored [116].

It is extremely important to investigate the accuracy and appropriateness of any proposed models using measured data of a large pool of subjects, with both patients and regular populations, to reach any conclusion on a PTT/PAT-based device. Different standards, such as AAMI/ ISO, ESH-IP [137], and BHS [138], have been used to validate such a BP device, but insufficient data and varied statistical tools made it inadequate for comparisons. Those reported validations, at times, compromised on the number of subjects and observers required and thus failed to prove that the newly-developed devices were comparable with medical-grade systems. Since PTT/PAT-based devices are calibration dependent, there is a dire need to include the accuracy of the reference device. However, the gold standard for BP measurement in clinical settings, namely the arterial invasive line, is usually applied for inpatients in serious medication conditions only, while the widely-used cuff-based devices are not error free [139]. Therefore, the unavailability of high-precision clinical data has limited the investigations and validations of those PTT/PAT-based BP devices. Further, it is not feasible to acquire invasive BP data of a large number of various populations (including healthy people) for correlation in order to obtain statistical significance. Last but not least, it is also necessary to evaluate wearable continuous cuffless BP against continuous arterial (invasive/non-invasive) BP. Since, these PTT/PAT-based BP devices aim to provide continuous data, it would be more relevant to be validated/calibrated using continuous beat-to-beat BP, rather than the BP averaged over a time, or discrete BP values.

In general, PTT/PAT-based BP devices have the capability of providing home-based monitoring in the form of a wearable linked with a smart device and thus mobile-heath (m-health) could be achieved with the connection to a cloud-based server. As a result, various approaches, such as the device with three chest sensors for PPG, ECG and ICG (impedance cardiogram) underneath a t-shirt, or ECG and PPG monitoring using circuitry on the toilet seats, beds and steering wheels, have been proposed and implemented [19,20,128,140]. In our group, the replacement of the three-electrode ECG configuration with a non-contact electrode (NCE) approach for ECG acquisition has been attempted. Different from regular contact-electrode ECG approaches which are dependent on the electrode-skin interface, NCE ECG could be obtained without any effects from the skin (sweat, hair, etc.), thus becoming of interest for off-the-clinic measurements. This method can further ease the signal acquisition as ECG data can be recorded using a single capacitive-coupling NCE in a specific point within the human body (i.e., wrist), thus holding promise for the realization of unobtrusive home-based BP monitoring wearable devices [38]. For instance, this approach can enable BP monitoring during sleeping or exercising without any hindrance to the users. Wireless communication via Bluetooth Low Energy (BLE) can be utilized to facilitate data collection and transfer locally among wearables for one user and globally with a cloud-based server, paving the way for real-time monitoring and diagnoses as well as distanced- and self-care. However, the communication time among devices needs to be considered as it affects directly the calculation of the "Time Delay" [141].

5. Conclusions

With the present technology, it is possible to implement a PTT/PAT-based system that accurately predicts the trends of BP instead of measuring BP itself. Considerable fluctuations in this trend can be used as a warning signal for users to monitor their BP and continuous monitoring of this variation can be helpful in clinical environments, as most vital parameters in operational theatres are measured continuously except for BP. Therefore, the development of a cuffless BP monitoring system will provide novel solutions in various medical scenarios. The frequency of calibration as predicted by many studies was less than one hour, which is inappropriate as the artery stiffness and dimension could not change abruptly. This indicates that confounding factors need to be taken into account for PTT/PAT-based BP as many other factors apart from the "Time Delay" also have their contributions. From comparisons, we found that the inclusion of heart rate improved the efficiency of PAT-based BP measurement. Further, one may ask whether there is a need for continuous monitoring of BP, and the answer may be not for healthy populations but crucial for CVD patients. Sleep apnea patients also require continuous BP monitoring; thus, a home-based system with connectivity to the caregiver online network would be of interest.

Acknowledgments: Manuja Sharma (M.S.) is supported by the scholarship under the Washington Research Foundation and the NSF CAREER Grant #1652818 under Hung Cao (H.C.).

Author Contributions: Manuja Sharma, Victor Ho, Devon Griggs and Karinne Barbosa carried out experiments. Manuja Sharma processed and analyzed the data. Tadesse Ghirmai, Sandeep K. Krishnan, Tzung K. Hsiai and Jung-Chih Chiao gave advises and discussions. Manuja Sharma and Hung Cao initiated the manuscript. Hung Cao supervised and sponsored the entire work. All authors read the manuscript.

Conflicts of Interest: The authors declare no conflict of interest.

References

1. Laflamme, M.A.; Murry, C.E. Heart regeneration. *Nature* **2011**, *473*, 326–335. [CrossRef] [PubMed]
2. Cao, H.; Kang, B.J.; Lee, C.-A.; Shung, K.K.; Hsiai, T.K. Electrical and mechanical strategies to enable cardiac repair and regeneration. *IEEE Rev. Biomed. Eng.* **2015**, *8*, 114–124. [CrossRef] [PubMed]
3. Centers of Disease Control and Prevention. Vital signs: Awareness and treatment of uncontrolled hypertension among adults—United States, 2003–2010. *Morb. Mortal. Wkly. Rep.* **2012**, *61*, 703–709.
4. Ubolsakka-Jones, C.; Sangthong, B.; Aueyingsak, S.; Jones, D.A. Older Women with Controlled Isolated Systolic Hypertension: Exercise and Blood Pressure. *Med. Sci. Sports Exerc.* **2016**, *48*, 983–989. [CrossRef] [PubMed]
5. Moroz, L.A.; Simpson, L.L.; Rochelson, B. Management of severe hypertension in pregnancy. *Semin. Perinatol.* **2016**, *40*, 112–118. [CrossRef] [PubMed]
6. Booth, J. A short history of blood pressure measurement. *Proc. R. Soc. Med.* **1977**, *70*, 793–799. [PubMed]
7. Rocci, S.R. The technique of sphygmomanometry. *Gazz. Med. Torino* **1897**, *10*, 981–1017.
8. Korotkoff, N. On methods of studying blood pressure. *Bull. Imp. Mil. Med. Acad.* **1905**, *11*, 365–367.
9. Perloff, D.; Grim, C.; Flack, J.; Frohlich, E.D.; Hill, M.; McDonald, M.; Morgenstern, B.Z. Human blood pressure determination by sphygmomanometry. *Circulation* **1993**, *88*, 2460–2470. [CrossRef] [PubMed]
10. Rivarocci, S. Un nuovo sfigmomanometro. *Gazz. Med. Torino* **1896**, *47*, 981–1017.

11. O'Brien, E.; Atkins, N. Accuracy of an oscillometric automatic blood pressure device: The Omron HEM403C. *J. Hum. Hypertens.* **1995**, *9*, 169–174.

12. O'Brien, E.; Waeber, B.; Parati, G.; Staessen, J. Blood pressure measuring devices: recommendations of the European Society of Hypertension. *Br. Med. J.* **2001**, *322*, 531–536. [CrossRef]

13. Leung, A.A.; Nerenberg, K.; Daskalopoulou, S.S.; McBrien, K.; Zarnke, K.B.; Dasgupta, K.; Cloutier, L.; Gelfer, M.; Lamarre-Cliché, M.; Milot, A.; et al. Hypertension Canada's 2016 Canadian Hypertension Education Program Guidelines for blood pressure measurement, diagnosis, assessment of risk, prevention, and treatment of hypertension. *Can. J. Cardiol.* **2016**, *32*, 569–588. [CrossRef] [PubMed]

14. JCS Joint Working Group. Guidelines for the clinical use of 24 h ambulatory blood pressure monitoring (ABPM) (JCS 2010). *Circ. J.* **2012**, *76*, 508–519.

15. Porter, K.B.; O'Brien, W.F.; Kiefert, V.; Knuppel, R.A. Flnapres: A Noninvasive Device To Monitor Blood Pressure. *Obstet. Gynecol.* **1991**, *78*, 430–433. [PubMed]

16. Kermode, J.; Davis, N.; Thompson, W. Comparison of the Finapres blood pressure monitor with intra-arterial manometry during induction of anaesthesia. *Anaesth. Intensive Care* **1989**, *17*, 470–475. [PubMed]

17. Thomas, S.S.; Nathan, V.; Zong, C.; Akinbola, E.; Aroul, A.L.P.; Philipose, L.; Soundarapandian, K.; Shi, X.; Jafari, R. BioWatch—A wrist watch based signal acquisition system for physiological signals including blood pressure. In Proceedings of the 2014 36th Annual International Conference of the IEEE Engineering in Medicine and Biology Society (EMBC), Chicago, IL, USA, 26–30 August 2014; pp. 2286–2289.

18. Kim, J.; Park, J.; Kim, K.; Chee, Y.; Lim, Y.; Park, K. Development of a nonintrusive blood pressure estimation system for computer users. *Telemed. E-Health* **2007**, *13*, 57–64. [CrossRef] [PubMed]

19. Kim, J.S.; Chee, Y.J.; Park, J.W.; Choi, J.W.; Park, K.S. A new approach for non-intrusive monitoring of blood pressure on a toilet seat. *Physiol. Meas.* **2006**, *27*, 203–211. [CrossRef] [PubMed]

20. Baek, H.J.; Lee, H.B.; Kim, J.S.; Choi, J.M.; Kim, K.K.; Park, K.S. Nonintrusive biological signal monitoring in a car to evaluate a driver's stress and health state. *Telemed. E-Health* **2009**, *15*, 182–189. [CrossRef] [PubMed]

21. Gu, W.; Poon, C.; Leung, H.; Sy, M.; Wong, M.; Zhang, Y. A novel method for the contactless and continuous measurement of arterial blood pressure on a sleeping bed. In Proceedings of the 2009 Annual International Conference of the IEEE Engineering in Medicine and Biology Society(EMBC), Minneapolis, MN, USA, 3–6 September 2009; pp. 6084–6086.

22. Nye, E. The effect of blood pressure alteration on the pulse wave velocity. *Br. Heart J.* **1964**, *26*, 261–265. [CrossRef] [PubMed]

23. Gribbin, B.; Steptoe, A.; Sleight, P. Pulse wave velocity as a measure of blood pressure change. *Psychophysiology* **1976**, *13*, 86–90. [CrossRef] [PubMed]

24. Ahmad, S.; Chen, S.; Soueidan, K.; Batkin, I.; Bolic, M.; Dajani, H.; Groza, V. Electrocardiogram-assisted blood pressure estimation. *IEEE Trans. Biomed. Eng.* **2012**, *59*, 608–618. [CrossRef] [PubMed]

25. Choi, Y.; Zhang, Q.; Ko, S. Noninvasive cuffless blood pressure estimation using pulse transit time and Hilbert–Huang transform. *Comput. Electr. Eng.* **2013**, *39*, 103–111. [CrossRef]

26. Jeong, I.C.; Wood, J.; Finkelstein, J. Using individualized pulse transit time calibration to monitor blood pressure during exercise. *Inf. Manag. Technol. Healthc.* **2013**, *190*, 39–41.

27. Yoon, Y.; Cho, J.H.; Yoon, G. Non-constrained blood pressure monitoring using ECG and PPG for personal healthcare. *J. Med. Syst.* **2009**, *33*, 261–266. [CrossRef] [PubMed]

28. Špulák, D.; Čmejla, R.; Fabián, V. Parameters for mean blood pressure estimation based on electrocardiography and photoplethysmography. In Proceedings of the 2011 International Conference on Applied Electronics (AE), Pilsen, Czech Republic, 7–8 September 2011; pp. 1–4.

29. Lass, J.; Meigas, I.; Karai, D.; Kattai, R.; Kaik, J.; Rossmann, M. Continuous blood pressure monitoring during exercise using pulse wave transit time measurement. In Proceedings of the 26th Annual International Conference of the IEEE Engineering in Medicine and Biology Society (IEMBS 2004), San Francisco, CA, USA, 1–4 September 2004; pp. 2239–2242.

30. Poon, C.; Zhang, Y. Cuff-less and noninvasive measurements of arterial blood pressure by pulse transit time. In Proceedings of the 27th Annual International Conference of the Engineering in Medicine and Biology Society (EMBS 2005), Shanghai, China, 1–4 September 2005; pp. 5877–5880.

31. Lobodzinski, S.S.; Laks, M.M. New devices for very long-term ECG monitoring. *Cardiol. J.* **2012**, *19*, 210–214. [CrossRef] [PubMed]

32. Cattivelli, F.S.; Garudadri, H. Noninvasive cuffless estimation of blood pressure from pulse arrival time and heart rate with adaptive calibration. In Proceedings of the 2009 Sixth International Workshop on Wearable and Implantable Body Sensor Networks, Berkeley, CA, USA, 3–5 June 2009; pp. 114–119.

33. McCarthy, B.; O'Flynn, B.; Mathewson, A. An investigation of pulse transit time as a non-invasive blood pressure measurement method. *J. Phys. Conf. Ser.* **2011**, *307*, 012060. [CrossRef]

34. Singh, R.B.; Cornélissen, G.; Weydahl, A.; Schwartzkopff, O.; Katinas, G.; Otsuka, K.; Watanabe, Y.; Yano, S.; Mori, H.; Ichimaru, Y.; et al. Circadian heart rate and blood pressure variability considered for research and patient care. *Int. J. Cardiol.* **2003**, *87*, 9–28. [CrossRef]

35. Liu, Q.; Yan, B.P.; Yu, C.-M.; Zhang, Y.-T.; Poon, C.C. Attenuation of systolic blood pressure and pulse transit time hysteresis during exercise and recovery in cardiovascular patients. *IEEE Trans. Biomed. Eng.* **2014**, *61*, 346–352. [PubMed]

36. Wong, Y.; Zhang, Y. The effects of exercises on the relationship between pulse transit time and arterial blood pressure. In Proceedings of the 27th Annual International Conference of the Engineering in Medicine and Biology Society (EMBS 2005), Shanghai, China, 1–4 September 2005; pp. 5576–5578.

37. Wong, M.Y.-M.; Pickwell-MacPherson, E.; Zhang, Y.-T. The acute effects of running on blood pressure estimation using pulse transit time in normotensive subjects. *Eur. J. Appl. Physiol.* **2009**, *107*, 169–175. [CrossRef] [PubMed]

38. Griggs, D.; Sharma, M.; Naghibi, A.; Wallin, C.; Ho, V.; Barbosa, K.; Ghirmai, T.; Cao, H.; Krishnan, S.K. Design and development of continuous cuff-less blood pressure monitoring devices. In Proceedings of the 2016 IEEE Sensors, Orlando, FL, USA, 30 October–2 November 2016; pp. 1–3.

39. Goldberger, A.L.; Amaral, L.A.; Glass, L.; Hausdorff, J.M.; Ivanov, P.C.; Mark, R.G.; Mietus, J.E.; Moody, G.B.; Peng, C.-K.; Stanley, H.E. Physiobank, physiotoolkit, and physionet components of a new research resource for complex physiologic signals. *Circulation* **2000**, *101*, e215–e220. [CrossRef] [PubMed]

40. Fye, W.B. A history of the origin, evolution, and impact of electrocardiography. *Am. J. Cardiol.* **1994**, *73*, 937–949. [CrossRef]

41. Sundnes, J.; Lines, G.T.; Cai, X.; Nielsen, B.F.; Mardal, K.-A.; Tveito, A. *Computing the Electrical Activity in the Heart*; Springer Science & Business Media: Berlin, Germany, 2007; Volume 1.

42. Braunwald, E.; Isselbacher, K.J.; Wilson, J.D.; Martin, J.B.; Kasper, D.; Hauser, S.L.; Longo, D.L. *Harrison's Principles of Internal Medicine*, 14th ed.; McGraw-Hill: New York, NY, USA, 1997.

43. Alastruey, J.; Parker, K.H.; Sherwin, S.J. Arterial pulse wave haemodynamics. In Proceedings of the 11th International Conference on Pressure Surges, Lisbon, Portugal, 24–26 October 2012; pp. 401–442.

44. Nichols, W.W.; O'Rourke, M.F.; Kenney, W.L. McDonald's Blood Flow in Arteries: Theoretical, Experimental and Clinical Principles, 3rd ed. *J. Cardiopulm. Rehabil. Prev.* **1991**, *11*, 407. [CrossRef]

45. Penaz, J. Photoelectric measurement of blood pressure, volume and flow in the finger. In *Digest of the 10th International Conference on Medical and Biological Engineering*; The Conference Committee: Dresden, Germany, 1973.

46. Wagenseil, J.E.; Mecham, R.P. Elastin in large artery stiffness and hypertension. *J. Cardiovasc. Transl. Res.* **2012**, *5*, 264–273. [CrossRef] [PubMed]

47. Geddes, L.A. *Handbook of Blood Pressure Measurement*; Springer Science & Business Media: Berlin, Germany, 1991.

48. Peter, L.; Noury, N.; Cerny, M. A review of methods for non-invasive and continuous blood pressure monitoring: Pulse transit time method is promising? *IRBM* **2014**, *35*, 271–282. [CrossRef]

49. Hughes, D.; Babbs, C.; Geddes, L.; Bourland, J. Measurements of Young's modulus of elasticity of the canine aorta with ultrasound. *Ultrason. Imaging* **1979**, *1*, 356–367. [CrossRef] [PubMed]

50. Harkness, M.L.; Harkness, R.; McDonald, D. The collagen and elastin content of the arterial wall in the dog. *Proc. R. Soc. Lond. B Biol. Sci.* **1957**, *146*, 541–551. [CrossRef] [PubMed]

51. Burton, A.C. Relation of structure to function of the tissues of the wall of blood vessels. *Physiol. Rev.* **1954**, *34*, 619–642. [PubMed]

52. Ferrari, A.U.; Radaelli, A.; Centola, M. Invited review: aging and the cardiovascular system. *J. Appl. Physiol.* **2003**, *95*, 2591–2597. [CrossRef] [PubMed]

53. O'rourke, M.F.; Hashimoto, J. Mechanical factors in arterial aging: A clinical perspective. *J. Am. Coll. Cardiol.* **2007**, *50*, 1–13. [CrossRef] [PubMed]

54. Mukkamala, R.; Hahn, J.-O.; Inan, O.T.; Mestha, L.K.; Kim, C.-S.; Töreyin, H.; Kyal, S. Toward ubiquitous blood pressure monitoring via pulse transit time: Theory and practice. *IEEE Trans. Biomed. Eng.* **2015**, *62*, 1879–1901. [CrossRef] [PubMed]

55. Cox, R.H. Regional variation of series elasticity in canine arterial smooth muscles. *Am. J. Physiol. Heart Circ. Physiol.* **1978**, *234*, H542–H551.

56. Moens, A.I. *Die Pulskurve*; EJ Brill: Leiden, The Netherlands, 1878.

57. Kanda, T.; Nakamura, E.; Moritani, T.; Yamori, Y. Arterial pulse wave velocity and risk factors for peripheral vascular disease. *Eur. J. Appl. Physiol.* **2000**, *82*, 1–7. [CrossRef] [PubMed]

58. Smith, R.P.; Argod, J.; Pépin, J.-L.; Lévy, P.A. Pulse transit time: an appraisal of potential clinical applications. *Thorax* **1999**, *54*, 452–457. [CrossRef] [PubMed]

59. Loukogeorgakis, S.; Dawson, R.; Phillips, N.; Martyn, C.N.; Greenwald, S.E. Validation of a device to measure arterial pulse wave velocity by a photoplethysmographic method. *Physiol. Meas.* **2002**, *23*, 581. [CrossRef] [PubMed]

60. Allen, J.; Murray, A. Age-related changes in peripheral pulse timing characteristics at the ears, fingers and toes. *J. Hum. Hypertens.* **2002**, *16*, 711–717. [CrossRef] [PubMed]

61. Chen, Y.; Wen, C.; Tao, G.; Bi, M.; Li, G. Continuous and noninvasive blood pressure measurement: A novel modeling methodology of the relationship between blood pressure and pulse wave velocity. *Ann. Biomed. Eng.* **2009**, *37*, 2222–2233. [CrossRef] [PubMed]

62. Nitzan, M.; Khanokh, B.; Slovik, Y. The difference in pulse transit time to the toe and finger measured by photoplethysmography. *Physiol. Meas.* **2001**, *23*, 85–93. [CrossRef]

63. Chen, Y.; Wen, C.; Tao, G.; Bi, M. Continuous and noninvasive measurement of systolic and diastolic blood pressure by one mathematical model with the same model parameters and two separate pulse wave velocities. *Ann. Biomed. Eng.* **2012**, *40*, 871–882. [CrossRef] [PubMed]

64. Allen, J. Photoplethysmography and its application in clinical physiological measurement. *Physiol. Meas.* **2007**, *28*, R1–R39. [CrossRef] [PubMed]

65. Chen, M.W.; Kobayashi, T.; Ichikawa, S.; Takeuchi, Y.; Togawa, T. Continuous estimation of systolic blood pressure using the pulse arrival time and intermittent calibration. *Med. Biol. Eng. Comput.* **2000**, *38*, 569–574. [CrossRef] [PubMed]

66. Zheng, Y.; Poon, C.C.; Zhang, Y.-T. Investigation of temporal relationship between cardiovascular variables for cuffless blood pressure estimation. In Proceedings of the 2012 IEEE-EMBS International Conference on Biomedical and Health Informatics (BHI), Hong Kong, China, 5–7 January 2012; pp. 644–646.

67. Ma, T.; Zhang, Y. A correlation study on the variabilities in pulse transit time, blood pressure, and heart rate recorded simultaneously from healthy subjects. In Proceedings of the 27th Annual International Conference of the Engineering in Medicine and Biology Society (EMBS 2005), Shanghai, China, 1–4 September 2005; pp. 996–999.

68. Schneider, J.A.; Davidson, D.M.; Winchester, M.A.; Taylor, C.B. The covariation of blood pressure and pulse transit time in hypertensive patients. *Psychophysiology* **1981**, *18*, 301–306.

69. Ma, H.T. A blood pressure monitoring method for stroke management. *Biomed. Res. Int.* **2014**, *2014*, 571623. [CrossRef] [PubMed]

70. García, M.T.G.; Acevedo, M.F.T.; Guzmán, M.R.; de Montaner, R.A.; Fernández, B.F.; del Río Camacho, G.; González-Mangado, N. Puede ser el tiempo de tránsito de pulso útil para detectar hipertensión arterial en pacientes remitidos a la unidad de sueño? *Arch. Bronconeumol.* **2014**, *50*, 278–284. [CrossRef] [PubMed]

71. Forouzanfar, M.; Ahmad, S.; Batkin, I.; Dajani, H.R.; Groza, V.Z.; Bolic, M. Model-based mean arterial pressure estimation using simultaneous electrocardiogram and oscillometric blood pressure measurements. *IEEE Trans. Instrum. Meas.* **2015**, *64*, 2443–2452. [CrossRef]

72. Li, Q.; Belz, G. Systolic time intervals in clinical pharmacology. *Eur. J. Clin. Pharmacol.* **1993**, *44*, 415–421. [CrossRef] [PubMed]

73. Fung, P.; Dumont, G.; Ries, C.; Mott, C.; Ansermino, M. Continuous noninvasive blood pressure measurement by pulse transit time. *Conf. Proc. IEEE Eng. Med. Biol. Soc.* **2004**, *1*, 738–741. [PubMed]

74. Ochiai, R.; Takeda, J.; Hosaka, H.; Sugo, Y.; Tanaka, R.; Soma, T. The relationship between modified pulse wave transit time and cardiovascular changes in isoflurane anesthetized dogs. *J. Clin. Monit. Comput.* **1999**, *15*, 493–501. [CrossRef] [PubMed]

75. Solà, J.; Rimoldi, S.F.; Allemann, Y. *Ambulatory Monitoring of the Cardiovascular System: The Role of Pulse Wave Velocity*; INTECH Open Access Publisher: Rijeka, Croatia, 2010.

76. Naschitz, J.E.; Bezobchuk, S.; Mussafia-Priselac, R.; Sundick, S.; Dreyfuss, D.; Khorshidi, I.; Karidis, A.; Manor, H.; Nagar, M.; Peck, E.R.; et al. Pulse transit time by R-wave-gated infrared photoplethysmography: Review of the literature and personal experience. *J. Clin. Monit. Comput.* **2004**, *18*, 333–342. [CrossRef] [PubMed]

77. Payne, R.; Symeonides, C.; Webb, D.; Maxwell, S. Pulse transit time measured from the ECG: An unreliable marker of beat-to-beat blood pressure. *J. Appl. Physiol.* **2006**, *100*, 136–141. [CrossRef] [PubMed]

78. Young, C.C.; Mark, J.B.; White, W.; DeBree, A.; Vender, J.S.; Fleming, A. Clinical evaluation of continuous noninvasive blood pressure monitoring: accuracy and tracking capabilities. *J. Clin. Monit.* **1995**, *11*, 245–252. [CrossRef] [PubMed]

79. Zhang, G.; Gao, M.; Xu, D.; Olivier, N.B.; Mukkamala, R. Pulse arrival time is not an adequate surrogate for pulse transit time as a marker of blood pressure. *J. Appl. Physiol.* **2011**, *111*, 1681–1686. [CrossRef] [PubMed]

80. Noordergraaf, A. *Circulatory System Dynamics*; Elsevier: Amsterdam, The Netherlands, 2012; Volume 1.

81. Marie, G.V.; Lo, C.; van Jones, J.; Johnston, D.W. The relationship between arterial blood pressure and pulse transit time during dynamic and static exercise. *Psychophysiology* **1984**, *21*, 521–527. [CrossRef] [PubMed]

82. Proença, J.; Muehlsteff, J.; Aubert, X.; Carvalho, P. Is pulse transit time a good indicator of blood pressure changes during short physical exercise in a young population? In Proceedings of the 2010 Annual International Conference of the IEEE Engineering in Medicine and Biology (EMBC), Buenos Aires, Argentina, 31 August–4 September 2010; pp. 598–601.

83. Wong, M.Y.-M.; Poon, C.C.-Y.; Zhang, Y.-T. An evaluation of the cuffless blood pressure estimation based on pulse transit time technique: A half year study on normotensive subjects. *Cardiovasc. Eng.* **2009**, *9*, 32–38. [CrossRef] [PubMed]

84. Baek, H.J.; Kim, K.K.; Kim, J.S.; Lee, B.; Park, K.S. Enhancing the estimation of blood pressure using pulse arrival time and two confounding factors. *Physiol. Meas.* **2009**, *31*, 145–157. [CrossRef] [PubMed]

85. Wibmer, T.; Doering, K.; Kropf-Sanchen, C.; Rüdiger, S.; Blanta, I.; Stoiber, K.; Rottbauer, W.; Schumann, C. Pulse transit time and blood pressure during cardiopulmonary exercise tests. *Physiol. Res.* **2014**, *63*, 287–296. [PubMed]

86. Masè, M.; Mattei, W.; Cucino, R.; Faes, L.; Nollo, G. Feasibility of cuff-free measurement of systolic and diastolic arterial blood pressure. *J. Electrocardiol.* **2011**, *44*, 201–207. [CrossRef] [PubMed]

87. Marcinkevics, Z.; Greve, M.; Aivars, J.I.; Erts, R.; Zehtabi, A.H. Relationship between arterial pressure and pulse wave velocity using photoplethysmography during the post-exercise recovery period. *Acta Univ. Latv. Biol.* **2009**, *753*, 59–68.

88. Jadooei, A.; Zaderykhin, O.; Shulgin, V. Adaptive algorithm for continuous monitoring of blood pressure using a pulse transit time. In Proceedings of the 2013 IEEE XXXIII International Scientific Conference Electronics and Nanotechnology (ELNANO), Kiev, Ukraine, 16–19 April 2013; pp. 297–301.

89. Gesche, H.; Grosskurth, D.; Küchler, G.; Patzak, A. Continuous blood pressure measurement by using the pulse transit time: comparison to a cuff-based method. *Eur. J. Appl. Physiol.* **2012**, *112*, 309–315. [CrossRef] [PubMed]

90. Ding, X.-R.; Zhang, Y.-T.; Liu, J.; Dai, W.-X.; Tsang, H.K. Continuous cuffless blood pressure estimation using pulse transit time and photoplethysmogram intensity ratio. *IEEE Trans. Biomed. Eng.* **2016**, *63*, 964–972. [CrossRef] [PubMed]

91. Association for the Advancement of Medical Instrumentatio. *American National Standard: Electronic or Automated Sphygmomanometers*; AAMI: Washington, DC, USA, 1987.

92. Baldridge, B.R.; Burgess, D.E.; Zimmerman, E.E.; Carroll, J.J.; Sprinkle, A.G.; Speakman, R.O.; Li, S.-G.; Brown, D.R.; Taylor, R.F.; Dworkin, S.; et al. Heart rate-arterial blood pressure relationship in conscious rat before vs. after spinal cord transection. *Am. J. Physiol. Regul. Integr. Comp. Physiol.* **2002**, *283*, R748–R756. [CrossRef] [PubMed]

93. Griggs, M.S.D.; Naghibi, A.; Barbosa, K.; Ho, V.; Wallin, C.; Ghirmai, T.; Krishnan, S.; Cao, H. Design and Development of Continuous Cuff-Less Blood Pressure Monitoring Devices. In Proceedings of the 2016 IEEE Sensors, Orlando, FL, USA, 30 October–2 November 2016.

94. Saeed, M.; Villarroel, M.; Reisner, A.T.; Clifford, G.; Lehman, L.-W.; Moody, G.; Heldt, T.; Kyaw, T.H.; Moody, B.; Mark, R.G. Multiparameter Intelligent Monitoring in Intensive Care II (MIMIC-II): A public-access intensive care unit database. *Crit. Care Med.* **2011**, *39*, 952–960. [CrossRef] [PubMed]
95. Tomson, J.; Lip, G.Y. Blood pressure demographics: nature or nurture...... genes or environment? *BMC Med.* **2005**, *3*, 3. [CrossRef] [PubMed]
96. Bland, J.M.; Altman, D.G. Measuring agreement in method comparison studies. *Stat. Methods Med. Res.* **1999**, *8*, 135–160. [CrossRef] [PubMed]
97. Bland, J.M.; Altman, D. Statistical methods for assessing agreement between two methods of clinical measurement. *Lancet* **1986**, *327*, 307–310. [CrossRef]
98. Bunce, C. Correlation, agreement, and Bland-Altman analysis: Statistical analysis of method comparison studies. *Am. J. Ophthalmol.* **2009**, *148*, 4–6. [CrossRef] [PubMed]
99. Fagard, R.H. Exercise characteristics and the blood pressure response to dynamic physical training. *Med. Sci. Sports Exerc.* **2001**, *33*, S484–S492. [CrossRef] [PubMed]
100. Longo, A.; Geiser, M.H.; Riva, C.E. Posture changes and subfoveal choroidal blood flow. *Investig. Ophthalmol. Vis. Sci.* **2004**, *45*, 546–551. [CrossRef]
101. Parati, G.; Casadei, R.; Groppelli, A.; di Rienzo, M.; Mancia, G. Comparison of finger and intra-arterial blood pressure monitoring at rest and during laboratory testing. *Hypertension* **1989**, *13*, 647–655. [CrossRef] [PubMed]
102. Ohlsson, O.; Henningsen, N. Blood pressure, cardiac output and systemic vascular resistance during rest, muscle work, cold pressure test and psychological stress. *Acta Med. Scand.* **1982**, *212*, 329–336. [CrossRef] [PubMed]
103. Peckerman, A.; Saab, P.G.; McCabe, P.M.; Skyler, J.S.; Winters, R.W.; Llabre, M.M.; Schneiderman, N. Blood pressure reactivity and perception of pain during the forehead cold pressor test. *Psychophysiology* **1991**, *28*, 485–495. [CrossRef] [PubMed]
104. Al'Absi, M.; Bongard, S.; Buchanan, T.; Pincomb, G.A.; Licinio, J.; Lovallo, W.R. Cardiovascular and neuroendocrine adjustment to public speaking and mental arithmetic stressors. *Psychophysiology* **1997**, *34*, 266–275. [CrossRef] [PubMed]
105. Agras, S.W.; Horne, M.; Taylor, B.C. Expectation and the blood-pressure-lowering effects of relaxation. *Psychosom. Med.* **1982**, *44*, 389–395. [CrossRef] [PubMed]
106. Davis, W.B.; Thaut, M.H. The influence of preferred relaxing music on measures of state anxiety, relaxation, and physiological responses. *J. Music Ther.* **1989**, *26*, 168–187. [CrossRef]
107. Steptoe, A.; Smulyan, H.; Gribbin, B. Pulse wave velocity and blood pressure change: Calibration and applications. *Psychophysiology* **1976**, *13*, 488–493. [CrossRef] [PubMed]
108. Matsumura, K.; Miura, K.; Takata, Y.; Kurokawa, H.; Kajiyama, M.; Abe, I.; Fujishima, M. Changes in blood pressure and heart rate variability during dental surgery. *Am. J. Hypertens.* **1998**, *11*, 1376–1380. [CrossRef]
109. Wiley, R.L.; Dunn, C.L.; Cox, R.H.; Hueppchen, N.A.; Scott, M.S. Isometric exercise training lowers resting blood pressure. *Med. Sci. Sports Exerc.* **1992**, *24*, 749–754. [CrossRef] [PubMed]
110. Petrofsky, J.S.; Lind, A.R. Aging, isometric strength and endurance, and cardiovascular responses to static effort. *J. Appl. Physiol.* **1975**, *38*, 91–95. [PubMed]
111. Petrofsky, J.S.; Burse, R.L.; Lind, A. Comparison of physiological responses of women and men to isometric exercise. *J. Appl. Physiol.* **1975**, *38*, 863–868. [PubMed]
112. McCarthy, B.; Vaughan, C.; O'Flynn, B.; Mathewson, A.; Mathúna, C.Ó. An examination of calibration intervals required for accurately tracking blood pressure using pulse transit time algorithms. *J. Hum. Hypertens.* **2013**, *27*, 744–750. [CrossRef] [PubMed]
113. Ding, X.-R.; Zhang, Y.-T. Photoplethysmogram intensity ratio: A potential indicator for improving the accuracy of PTT-based cuffless blood pressure estimation. In Proceedings of the 2015 37th Annual International Conference of the IEEE Engineering in Medicine and Biology Society (EMBC), Milan, Italy, 25–29 August 2015; pp. 398–401.
114. Kachuee, M.; Kiani, M.M.; Mohammadzade, H.; Shabany, M. Cuff-Less Blood Pressure Estimation Algorithms for Continuous Health-Care Monitoring. *IEEE Trans. Biomed. Eng.* **2017**, *64*, 859–869. [CrossRef] [PubMed]
115. Kim, J.S.; Kim, K.K.; Baek, H.J.; Park, K.S. Effect of confounding factors on blood pressure estimation using pulse arrival time. *Physiol. Meas.* **2008**, *29*, 615. [CrossRef] [PubMed]

116. Drinnan, M.J.; Allen, J.; Murray, A. Relation between heart rate and pulse transit time during paced respiration. *Physiol. Meas.* **2001**, *22*, 425–432. [CrossRef] [PubMed]

117. di Rienzo, M.; Parati, G.; Radaelli, A.; Castiglioni, P. Baroreflex contribution to blood pressure and heart rate oscillations: Time scales, time-variant characteristics and nonlinearities. *Philos. Trans. R. S. Lond. A Math. Phys. Eng. Sci.* **2009**, *367*, 1301–1318. [CrossRef] [PubMed]

118. Nichols, W.; O'Rourke, M. *McDonald's Blood Flow in Arteries*, 4th ed.; Edward Arnold: London, UK, 1998.

119. Shaltis, P.; Reisner, A.; Asada, H. A hydrostatic pressure approach to cuffless blood pressure monitoring. In Proceedings of the 26th Annual International Conference of the IEEE Engineering in Medicine and Biology Society, San Francisco, CA, USA, 1–4 September 2004; pp. 2173–2176.

120. Gaddum, N.; Alastruey, J.; Beerbaum, P.; Chowienczyk, P.; Schaeffter, T. A technical assessment of pulse wave velocity algorithms applied to non-invasive arterial waveforms. *Ann. Biomed. Eng.* **2013**, *41*, 2617–2629. [CrossRef] [PubMed]

121. Gaballa, M.A.; Jacob, C.T.; Raya, T.E.; Liu, J.; Simon, B.; Goldman, S. Large Artery Remodeling During Aging Biaxial Passive and Active Stiffness. *Hypertension* **1998**, *32*, 437–443. [CrossRef] [PubMed]

122. Mayor, S. Doctors are urged to measure blood pressure in both arms. *BMJ* **2016**, *353*, i2577. [CrossRef] [PubMed]

123. Eşer, İ.; Khorshid, L.; Güneş, Ü.Y.; Demir, Y. The effect of different body positions on blood pressure. *J. Clin. Nurs.* **2007**, *16*, 137–140. [CrossRef] [PubMed]

124. Ng, K.-G. Review of measurement methods and clinical validation studies of noninvasive blood pressure monitors: Accuracy requirements and protocol considerations for devices that require patient-specific calibration by a secondary method or device before use. *Blood Press. Monit.* **2011**, *16*, 291–303. [CrossRef] [PubMed]

125. Han, H.; Kim, M.-J.; Kim, J. Development of real-time motion artifact reduction algorithm for a wearable photoplethysmography. In Proceedings of the 2007 29th Annual International Conference of the IEEE Engineering in Medicine and Biology Society, Lyon, France, 23–26 August 2007; pp. 1538–1541.

126. Poh, M.-Z.; Swenson, N.C.; Picard, R.W. Motion-tolerant magnetic earring sensor and wireless earpiece for wearable photoplethysmography. *IEEE Trans. Inf. Technol. Biomed.* **2010**, *14*, 786–794. [CrossRef] [PubMed]

127. Lee, H.; Lee, J.; Jung, W.; Lee, G.-K. The periodic moving average filter for removing motion artifacts from PPG signals. *Int. J. Control Autom. Syst.* **2007**, *5*, 701–706.

128. Solà, J.; Proença, M.; Ferrario, D.; Porchet, J.-A.; Falhi, A.; Grossenbacher, O.; Allemann, Y.; Rimoldi, S.F.; Sartori, C. Noninvasive and nonocclusive blood pressure estimation via a chest sensor. *IEEE Trans. Biomed. Eng.* **2013**, *60*, 3505–3513. [CrossRef] [PubMed]

129. Teng, X.; Zhang, Y. The effect of applied sensor contact force on pulse transit time. *Physiol. Meas.* **2006**, *27*, 675–684. [CrossRef] [PubMed]

130. Teng, X.-F.; Zhang, Y.-T. Theoretical study on the effect of sensor contact force on pulse transit time. *IEEE Trans. Biomed. Eng.* **2007**, *54*, 1490–1498. [CrossRef] [PubMed]

131. Yu, C.; Liu, Z.; McKenna, T.; Reisner, A.T.; Reifman, J. A method for automatic identification of reliable heart rates calculated from ECG and PPG waveforms. *J. Am. Med. Inf. Associ.* **2006**, *13*, 309–320. [CrossRef] [PubMed]

132. Foo, J.Y.A. Use of independent component analysis to reduce motion artifact in pulse transit time measurement. *IEEE Signal Proc. Lett.* **2008**, *15*, 124–126. [CrossRef]

133. Shin, H.S.; Lee, C.; Lee, M. Adaptive threshold method for the peak detection of photoplethysmographic waveform. *Comput. Biol. Med.* **2009**, *39*, 1145–1152. [CrossRef] [PubMed]

134. McDuff, D.; Gontarek, S.; Picard, R.W. Remote detection of photoplethysmographic systolic and diastolic peaks using a digital camera. *IEEE Trans. Biomed. Eng.* **2014**, *61*, 2948–2954. [CrossRef] [PubMed]

135. McCombie, D.B. *Development of a Wearable Blood Pressure Monitor Using Adaptive Calibration of Peripheral Pulse Transit Time Measurements*; Massachusetts Institute of Technology: Cambridge, MA, USA, 2008.

136. Douniama, C.; Sauter, C.; Couronne, R. Blood pressure tracking capabilities of pulse transit times in different arterial segments: A clinical evaluation. In Proceedings of the 2009 36th Annual Computers in Cardiology Conference (CinC), Park City, UT, USA, 13–16 September 2009; pp. 201–204.

137. O'brien, E.; Pickering, T.; Asmar, R.; Myers, M.; Parati, G.; Staessen, J.; Mengden, T.; Imai, Y.; Waeber, B.; Palatini, P.; et al. Working Group on Blood Pressure Monitoring of the European Society of Hypertension International Protocol for validation of blood pressure measuring devices in adults. *Blood Press. Monit.* **2002**, *7*, 3–17. [CrossRef] [PubMed]

138. O'brien, E.; Petrie, J.; Little, W.; de Swiet, M.; Padfield, P.L.; Altma, D.G.; Bland, M.; Coats, A.; Atkins, N. Short report: An outline of the revised British Hypertension Society protocol for the evaluation of blood pressure measuring devices. *J. Hypertens.* **1993**, *11*, 677–679. [CrossRef] [PubMed]

139. Ribezzo, S.; Spina, E.; di Bartolomeo, S.; Sanson, G. Noninvasive techniques for blood pressure measurement are not a reliable alternative to direct measurement: A randomized crossover trial in ICU. *Sci. World J.* **2014**, *2014*, 363628. [CrossRef] [PubMed]

140. Zheng, Y.-L.; Ding, X.-R.; Poon, C.C.Y.; Lo, B.P.L.; Zhang, H.; Zhou, X.-L.; Yang, G.-Z.; Zhao, N.; Zhang, Y.T. Unobtrusive sensing and wearable devices for health informatics. *IEEE Trans. Biomed. Eng.* **2014**, *61*, 1538–1554. [CrossRef] [PubMed]

141. Chandrasekaran, V.; Dantu, R.; Jonnada, S.; Thiyagaraja, S.; Subbu, K.P. Cuffless differential blood pressure estimation using smart phones. *IEEE Trans. Biomed. Eng.* **2013**, *60*, 1080–1089. [CrossRef] [PubMed]

technologies

MDPI

Article

Communication Challenges in On-Body and Body-to-Body Wearable Wireless Networks—A Connectivity Perspective

Dhafer Ben Arbia [1,*], Muhammad Mahtab Alam [2], Yannick Le Moullec [2] and Elyes Ben Hamida [3]

[1] SERCOM Lab, Polytechnic School of Tunisia, University of Carthage, B.P. 743, 2078 La Marsa, Tunisia
[2] Thomas Johann Seebeck Department of Electronics, Tallinn University of Technology, Ehitajate tee 5, 19086 Tallinn, Estonia; muhammad.alam@ttu.ee (M.M.A.); yannick.lemoullec@ttu.ee (Y.L.M.)
[3] IRT SystemX, Building N3, 8 Avenue de la Vauve, CS 90070, 91127 Palaiseau, France; elyes.ben-hamida@irt-systemx.fr
* Correspondence: dhafera@qmic.com; Tel.: +974-5010-8593

Received: 15 May 2017; Accepted: 30 June 2017; Published: 6 July 2017

Abstract: Wearable wireless networks (WWNs) offer innovative ways to connect humans and/or objects anywhere, anytime, within an infinite variety of applications. WWNs include three levels of communications: on-body, body-to-body and off-body communication. Successful communication in on-body and body-to-body networks is often challenging due to ultra-low power consumption, processing and storage capabilities, which have a significant impact on the achievable throughput and packet reception ratio as well as latency. Consequently, all these factors make it difficult to opt for an appropriate technology to optimize communication performance, which predominantly depends on the given application. In particular, this work emphasizes the impact of coarse-grain factors (such as dynamic and diverse mobility, radio-link and signal propagation, interference management, data dissemination schemes, and routing approaches) directly affecting the communication performance in WWNs. Experiments have been performed on a real testbed to investigate the connectivity behavior on two wireless communication levels: on-body and body-to-body. It is concluded that by considering the impact of above-mentioned factors, the general perception of using specific technologies may not be correct. Indeed, for on-body communication, by using the IEEE 802.15.6 standard (which is specifically designed for on-body communication), it is observed that while operating at low transmission power under realistic conditions, the connectivity can be significantly low, thus, the transmission power has to be tuned carefully. Similarly, for body-to-body communication in an indoor environment, WiFi IEEE 802.11n also has a high threshold of end-to-end disconnections beyond two hops (approximatively 25 m). Therefore, these facts promote the use of novel technologies such as 802.11ac, NarrowBand-IoT (NB-IoT) etc. as possible candidates for body-to-body communications as a part of the Internet of humans concept.

Keywords: wearable wireless networks (WWNs); on-body networks (BANs); body-to-body networks (BBNs); connectivity; disaster relief and emergency applications

---

1. Introduction

Internet of humans (IoH) is a new paradigm in which wearable technology is emerging as a cutting-edge enabler. IoH is the concept of connecting, monitoring and recording human data with the Internet. Wearable technology is revolutionizing many applications, including health-care, sports and fitness, rescue and emergency management, augmented reality, fashion, and so on [1]. Recently, wearable technology revenue has greatly increased, passing from USD 2 billion in 2013 to more than USD 15 billion in 2017. Furthermore, technology ownership has undergone a strong increase, from 7% in 2014 to 14% by 2015. By the end of 2017, it is predicted to double again and reach 28% [2].

Nowadays people can easily keep track of their health and fitness with human-assistive wearable technology. Elderly people can also be remotely monitored and followed up. For instance, it was recently reported that the UK National Health Service (NHS) could save up to 7 billion pounds per year by using innovative technologies to deliver quality health-care to chronically ill with fewer hospital visits and admissions [1,3].

Virtual and augmented reality (VR/AR) form another showcase of wearable technology which has completely changed the perception of immersive vision. Within next few years, many millions of people will be able to walk around wearing relatively unobtrusive AR devices that offer an immersive and high-resolution view of a visually-augmented world [4]. Among other applications, wearable technology also hails to assist first responders in rescuing and evacuating people during disasters. In past few years, it has been found that wearable technology can be vigorously exploited in disaster contexts to not only save human lives but also to monitor the real-time health status of the rescue team members and victims. Furthermore, it helps operations commanders to make optimal decisions during disaster relief operations.

Other use cases within recent works targeted real testbeds and implementations in order to evaluate the performance of the wearable wireless networks (WWN) integrated with Internet of things (IoT) in real conditions. Miranda et al. in [5] implemented and evaluated a complete common recognition and identification platform (CRIP) for healthcare IoT. CRIP enables a basic configuration and communication standardization of healthcare "things". Other aspects are also covered, in particular security and privacy, and health device integration. Different communication standards were used to deploy CRIP, such as Near Field Communication (NFC), biometrics (fingerprints) and Bluetooth. In most of the above-mentioned applications, wireless communication is inevitable between various types of devices including sensors, actuators, coordinators, and gateways. Additionally, with the advent of body-to-body networks (BBNs or B2B), the communication is extended from classical "on-body networks/body area networks (BANs)" to modern "body-to-body networks (BBNs)" as shown in Figure 1. Consequently, wearable wireless networks (WWNs) are emerging as a new frontier for future smart applications in Internet of things (IoT) and Internet of humans (IoH). From the viewpoint of WWN "connectivity" in IoT and IoH, BBNd provide multi-hop device-to-device (D2D) communication to extend the end-to-end network coverage. This coincides with the vision of 5G, setting up new challenges towards cooperative and collaborative D2D communication among heterogeneous devices.

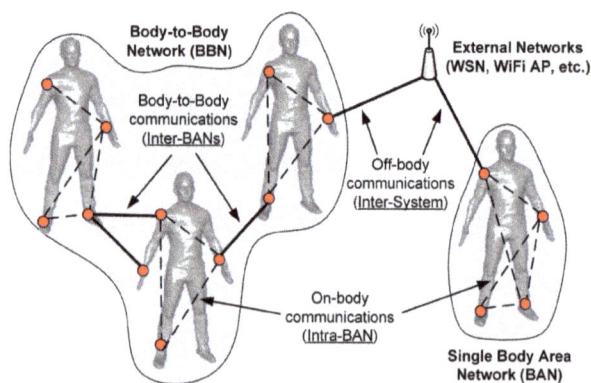

Figure 1. Wearable wireless networks: extending on-body communication to body-to-body and off-body communication. BAN: on-body network.

The "connectivity" between devices can not be merely ensured by establishing and executing a hand-shaking protocol. For example, if the packet reception ratio is as low as 50% it means that half of the time devices are disconnected, or in other words, the packets delay recorded is 2 times higher than the given application constraint. Additionally, it also means that only 50% of the time the considered devices are connected. Consequently, connectivity is an indispensable aspect in WWNs and IoT. To have deeper understanding in this article, we take a closer look at the BAN and BBN communication in WWNs. In particular, we accentuate on the network connectivity challenges in WWNs while considering a first responder rescue and critical operation as a case study.

The contributions of this paper are the following:

- *Potential technologies for BAN and BBN wireless networks for the considered application*: A comprehensive overview of existing technologies is presented with potential candidates for BAN and BBN communication. We consider various cross-layer design parameters that can have a direct or indirect impact on the connectivity. Most appropriate standards and supported technologies are selected for both BAN and BBN communication in the application context (use case: disaster and relief operations).
- *Impact of space-time channel variations, co-channel interference for on-body connectivity*: On-body connectivity is addressed from the view points of space–time channel variations and dynamic mobility as well as co-channel interference under the constraints of low power, latency and packet delivery ratio (PDR).
- *Real testbed to evaluate and analyze end-to-end connectivity and end-to-end round trip time delay for BBN wireless communication*: The analysis of the end-to-end connectivity and the end-to-end round trip time delay behavior for BBN communication is discussed with regards to the data dissemination strategies. Then, the experiment scenario is presented and the testing is detailed based on the newly proposed dedicated routing strategy for a disaster context [6] (Optimized Routing Approach for Critical and Emergency Networks (i.e., ORACE-Net)). This work is a rare implementation of BAN and BBN technologies in the disaster context. It does not only evaluate the considered implementation, but it also provides a strong guidance for possible future emerging candidates for wireless BAN and BBN technologies.

2. Overview of Candidate Technologies for Wearable Wireless Networks

While selecting the appropriate technologies for WWNs, there are a number of constraints to be considered. Based on a given application, often power consumption is required to be very low in order to maximize the lifetime of deployed nodes and the whole network. Typically for on-body communication, excessive power can result in additional interference [7] and therefore fine tuning of the transmission power is extremely important. The effective throughput (after adding all overheads) is another constraint directly related with the specific technology being used. In addition, packet latency, dynamic range, node density and network topology are few important constraints.

Below, we summarize possible optimal selection of the technologies for both BAN and BBN communication.

2.1. Overview of On-Body Communication Technologies and Existing Implementations

A holistic overview of (non-cellular) existing standards and technologies for WWN applications is presented in Table 1. For on-body networks, over the past decade a legacy has been incurred from the wireless sensor network-powered Zigbee and the IEEE 802.15.4 standard. Such BAN products and protocol designs remained very dominant by covering health-care-related applications such as patient monitoring in hospital wards and intensive care units. Few other variants of this standard (for example IEEE 802.15.4a, mainly used for wireless sensor networks for ultra-wide bands, and IEEE 802.15.4j, a modified physical layer of IEEE 802.15.4) focused on the medical body area network (MBAN) with frequency band between 2360 MHz to 2400 MHz (just before the congested narrow band spectrum). However, this is limited to medical-related applications where for instance dynamic mobility, space–time variations of the wireless channel cause less impact and therefore such specific features are not proposed in the standard.

The SmartBAN standard is proposed by the European Telecommunications Standards Institute (ETSI) to support health-care-related applications. However, the proposed BAN-specific features are not scalable for covering other non-medical applications [8]. The IEEE 802.15.6 standard is targeted for wide a range of applications for body area networks. It provides great flexibility, diversity in terms of features and WWN-specific provisions which are necessary to be exploited in the dynamic and emerging applications. The standard proposes BAN-specific channel models (which are very important to accurately model the system performance).

In terms of maximum achievable throughput at narrow-band, the IEEE 802.15.6 standard can reach up to 680 Kb/s, while operating at maximum frequency and highest modulation order by considering all the overheads of the Media Access Control (MAC) and Physical (PHY) layers [9]. This imposes limits to the use of the IEEE 802.15.6 standard in a few emerging applications. For example in applications such as augmented reality where transmission of high-rate audio and video are necessary, the IEEE 802.15.6 standard does not meet the requirements.

However, since the IEEE 802.15.6 standard provides ultra-low power consumption for both invasive and non-invasive devices while having key security features, it is expected to cover a wide range of applications with relatively low throughput requirements. Importantly, the great flexibility on the usage of multiple options at the PHY (i.e., human body communication, narrow-band, and ultra-wide-band) and at the MAC layer (scheduled access, beacon enabled/disabled, carrier sense multiple access with collision avoidance (CSMA/CA), polling and posting), make the IEEE 802.15.6 standard a viable option for on-body WWN. In this paper we use the IEEE 802.15.6 standard with applications covering the data rate up to 600 kbps. Further, we provide an in-depth overview of the standard focused on connectivity.

Table 1. Comparisons of the key enabling standards for wearable wireless networks [10].

	IEEE 802.11a/b/g/n (WiFi)	IEEE 802.15.1 (Bluetooth)	IEEE 802.15.1 (Bluetooth-LE)	IEEE 802.15.4 (Zigbee)	IEEE 802.15.4a (UWB)	IEEE 802.15.4j (MBAN)	IEEE 802.15.6 (BANs standard)
Modes of Operation	Adhoc, Infrastructure	Adhoc	Adhoc	Adhoc	Adhoc	Adhoc	Adhoc
Physical (PHY) Layers	NB	NB	NB	NB	UWB	NB	NB,UWB,HBC
Radio Frequencies (MHz)	2400, 5000	2400	2400	868/915, 2400	75–724, 3000–5000, 6000–10,000	2360–2490, 2390–2400	402–405, 420–450, 863–870, 902–928, 950–956, 2360–2400, 2400–2438.5
Power Consumption	High (~800 mW)	Medium (~100 mW)	low (~10 mW)	Low (~60 mW)	Low (~50 mW)	Low (~50 mW)	Ultra low (1 mW at 1 m distance)
Maximal Signal Rate	Up to 150 Mb/s	Up to 3 Mb/s	Up to 1 Mb/s	Up to 250 Kb/s	Up to 27.24 Mb/s	Up to 250 Kb/s	10 Kb/s to 10 Mb/s
Communication Range	Up to 250 m (802.11n)	100 m (class 1 device)	Up to 100 m	Up to 75 m	Up to 30 m	Up to 75 m	Up to 10 m (nominal ~2 m)
Networking Topology	Infrastructure based	Ad-hoc very small networks	Ad-hoc very small networks	Ad-hoc, Peer-to-Peer, Star, Mesh	Ad-hoc, Peer-to-Peer, Star, Mesh	Ad-hoc, Peer-to-Peer, Star	Intra-WBAN: 1/2-hop star. Inter-WBANs: non-standardized
Topology size	2007 devices for structured WiFi BSS	Up to 8 devices per Piconet	Up to 8 devices per Piconet	Up to 65,536 devices per network	Up to 65,536 devices per network	Up to 65,536 devices per network	Up to 256 devices per body, and up to 10 WBANs in 6 mt
Target Applications	Data Networks	Voice Links	Healthcare, fitness, beacons, security, etc.	Sensor networks, home automation, etc.	Short range and high data rates, localization, etc.	Short range Medical Body Area Networks	Body Centric application
Target BAN Architectures	Off-Body	On-Body	On-Body	Body-to-Body, Off-Body	Body-to-Body	On-Body	On-Body

* NB: Narrow Band, UWB: Ultra-wide-band, HBC: Human body communication.

2.2. Body-to-Body Technologies and Implementations

For BBN communications, the devices are supposed to communicate over relatively long distances and often are equipped with better battery lifetime in comparison to BAN sensors. Table 1 summarizes the BBN communication standards. The distance from one BAN to another is likely to exceed several hundred meters and therefore short-range and low-power technologies (as mentioned for on-body) are not suitable choices. Consequently, the dynamic range constraint eliminates some candidates for BBN communications including IEEE 802.15.6, IEEE 802.15.4 (ZigBee), IEEE 802.15.4j Medical Body Area Network (MBAN) and IEEE 802.15.4a (ultra-wideband, UWB). As can be seen in Table 1, the eligible standards that can fulfill technical and operational requirements for BBN communications are; WiFi IEEE 802.11a/b/g/n and Bluetooth IEEE 802.15.1 (and Bluetooth low energy, BLE). In comparison to Bluetooth, WiFi is the most relevant to BBN communication for the following reasons: (1) the communication range of WiFi, which is up to 250 m (outdoor), is higher than the range assured by Bluetooth which could reach a maximum of 100 m (under specific conditions for class 1); (2) there are a large number of mobile devices implementing at least one of the WiFi varieties cited above; and (3) bandwidth assured by WiFi is around 150 Mb/s (in case of IEEE 802.11n) and could even reach 500 Mb/s (IEEE 802.11ac), compared to the hundreds of Kb/s offered by Bluetooth.

Furthermore, Bluetooth low energy (BLE) and WiFi are expected to provide short range coverage from 100 to 200 m with a throughput range from a few Mbps to hundreds of Mbps. BLE or Bluetooth Smart technologies are strong candidates for the BBN communications, however, they remain limited in terms of range (100 m theoretical) as well as low transmission power and therefore they are high power interference-sensitive. On the other hand, recent variants of the WiFi standard (i.e., IEEE 802.11n, IEEE 802.11ac, etc.) offer provisions to operate in multi-band frequencies (i.e., 2.4 GHz and 5 GHz). Additionally, using WiFi, devices could operate for more than 10 h with one battery. Thus, WiFi could be considered as a pertinent candidate for future BBN communications.

Indeed, these above statements were proved through recent extensive simulations in [11], where it was concluded that WiFi IEEE 802.11 has the best performance in the considered application (disaster relief networks) compared to ZigBee and Wireless Body Area Network (WBAN). Furthermore, a recent experiment [12] validated WiFi IEEE 802.11n as a BBN communication protocol for the disaster relief applications despite the limitations detailed in Section 4. To sum up, WiFi IEEE 802.11 standard in general remains the most appropriate technology for BBN communications.

2.3. Overview of Key WWN Applications and Implementations

The wireless technologies given above are considered as current and prospective technologies which fulfill the BAN and BBN communications. Wireless technology is selected and implemented depending on the requirements of the applications and use cases. Chen et al. in [13] classify the applications into three main classes: (1) remote health and fitness monitoring; (2) military and training; and (3) intelligent biosensors for vehicle-area-networks. Moreover, authors in [13] discuss a list of research projects and implementations, in particular: Advanced Health and Disaster Aid Network (AID-N) [14] targets disaster and public safety applications. AID-N uses wired connection for BAN communication, mesh and ZigBee for BBN. Off-body communication in AID-N are fulfilled through WiFi, cellular networks and the Internet. AID-N aims to sense pulse, blood pressure, temperature and Electrocardiography (ECG). Negra et al. in [15] focus more on the main medical applications: (1) telemedicine and remote patient monitoring; (2) rehabilitation and therapy; (3) biofeedback; and (4) ambient assisted living. The latter work discusses also the Quality of Service (QoS) requirements for the medical context. The earliest proposed schemes target to enhance the on-body devices transmission reliability and improve energy efficiency. Chen et al. in [16] proposed a novel cross-layer design optimization (CLDO) scheme. Indeed, the design of CLDO relies on the three lower layers (i.e., PHY, MAC and network layer). Power consumption is firstly optimized by selecting optimal power relays. Then, the remaining energy in leaf nodes is utilized to increase the lifetime and the reliability. An optimal packet size is given for energy efficiency. Chen et al. claim that an inevitably slight overhead accompanies CLDO processing for different factors. First, during network initialization complex procedures are run. Second, the algorithm uses a certain number of iterations which influences the overall performance. Third, CLDO lacks the capacity to manage dynamic location situations.

Recent optimization of existing approaches has been proposed to draw a mathematical model for joint single path routing and relay deployment in BAN design. An interesting algorithm has been introduced by D'Andreagiovanni et al. in [17] to handle the uncertainty that affects traffic demands in the multiperiod capacitated network design problem (MP-CNDP). Additionally, a hybrid primal heuristic based on the combination of a randomized fixing algorithm was proposed by the authors, inspired from ant colony optimization and exact large neighborhood search. Performance of the proposed model has been confirmed based on computational experiment compared to existing solutions. This strategy has been improved by D'Andreagiovanni et al. in [18]. The authors adopt a best performance solution based on a min–max approach [19]. Indeed, the proposed algorithm relies on a combination of a probabilistic fixing procedure, guided by linear relaxations, and an exact large variable neighborhood search [17]. This combination has been inspired by the solution methods approximate nondeterministic tree-search (ANTS) [20], ant colony optimization [21] and other randomized algorithms [22]. D'Andreagiovanni et al. extended their preliminary work [18] by new integer linear programming (ILP) heuristic to solve the design problem. The new techniques detailed in [23] do not only fix the variables expressing routing decisions, but also employ an initial deterministic fixing phase of the variables modeling the activation of relay nodes. Experiment conducted by this work shows that the proposed approach outperforms the existing optimization solvers strategies and the results recorded in [18]. In [24], a heuristic min–max regret approach has been developed for BAN design, showing a significant reduction in the conservatism of optimal solutions with respect to the pure min–max approach of [18].

The main challenges in WWN are around routing techniques for BAN and BBN networks. We have recently proposed a new routing approach (i.e., ORACE-Net) which is dedicated to disaster and critical emergency networks. ORACE-Net [6] relies on end-to-end link quality estimation for routing decisions. The scope of this work is to present the network connectivity analysis of our proposed approach [6]. Another approach presented by Tsouri et al. in [25] relies on Dijkstra's algorithm augmented with novel link cost function designed to balance energy consumption across the network. This latter technique avoids relaying through nodes which spend more accumulated energy than others. Indeed, routing decisions are made based on the energy optimization. Authors claim that the proposed approach increases the network lifetime by 40% with a slight raise of the energy consumed per bit. This work is limited because the main concern of an operational application is studying the BBN network connectivity and routing which consists of the only present backbone in case of operational and dynamic context.

3. On-Body Communication and Connectivity

Typically, the on-body communication architecture is composed of sensors (to obtain physiological data), actuators (to act on obtained observations) and a coordinator (to control and coordinate both on-body and beyond body networks). Sensors (i.e., biological sensors, environment sensors, location and position sensors, etc.) could be connected directly to the on-body coordinator, which is often considered as more powerful with longer lifetime batteries than deployed sensors. Successful connectivity between on-body nodes and coordinators are often constrained due to ultra-low power consumption, processing and storage, data throughput or packet reception ratio and latency. These fine-grain constraints are often impacted by coarse-grain factors such as dynamic and diverse mobility (e.g., for sports and fitness applications), radio-link and signal propagation (indoor, outdoor, underwater, during emergencies and disasters etc.), interference management (co-channel, inter-channel) and coexistence strategies (time-shared, channel-hopping, collaborative etc.), and data dissemination schemes, as well as routing approaches. Consequently, to successfully support diverse wearable applications, the vital impact of coarse-grain factors is extremely important to analyze. In this section such factors are explored and presented.

3.1. Space–Time Varying Radio-Links and Signal Propagation

Accurate channel modeling is a very active topic of research in both BAN and BBN. In particular, under dynamic and diverse mobility patterns, based on the positions of on-body sensors, space and time varying radio-links can severely affect the connectivity between sensors-coordinator communication.

Such space–time varying characteristics often provide higher degree of correlation as identified in [26]. Further, both short-term and long-term channel fading can be modeled to precise the path-loss factors from signal propagation. Moreover, body shadowing also has to be taken into account for accurate modeling of on-body and body-to-body links [27].

In the IEEE 802.15.6 standard, the proposed channel models are limited to stationary radio-links, which consequently requires space-time variations and accurate signal propagation enhancements. At the narrow band, both proposed channel models (i.e., CM3-A and CM3-B) are distance-dependent; the path-loss derived for those models from measurement campaigns was recently enhanced using bio-mechanical mobility and deterministic models [28]. As an example shown in Figure 2, for a space–time varying link such as "wrist-chest" the average peak-to-peak path-loss is 10-dBs higher than for IEEE 802.15.6 standard channel models and hence is more accurate. Realistic radio-link conditions and signal propagation are important to analyze the true connectivity between on-body sensors-coordinator communication.

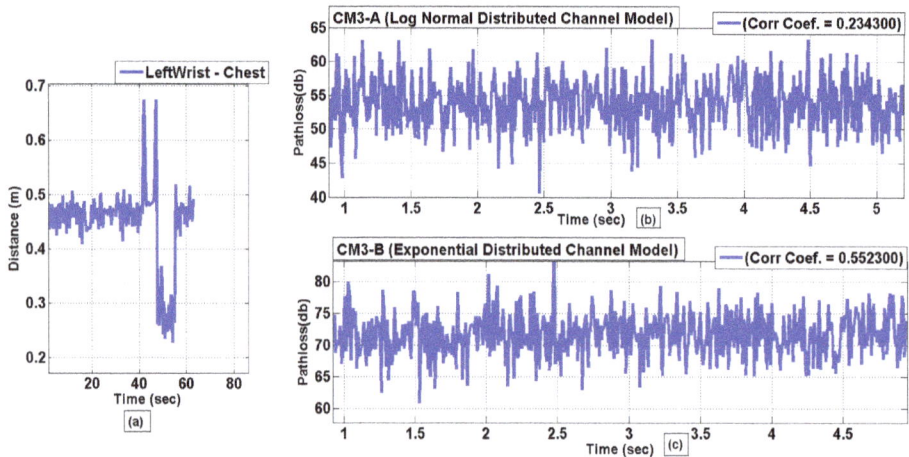

Figure 2. IEEE 802.15.6 enhanced path-loss models obtained from bio-mechanical deterministic channel model. For example, a link between left wrist and chest is shown. (**a**) Time-varying distances; (**b**) Enhanced pathloss model CM3-A and (**c**) enhanced pathloss model CM3-B.

Figure 3 shows the average packet delivery ratio (PDR) results of the on-body communication. Twelve sensors placed on various locations around the body include links which provide space–time variations, static as well as periodic line-of-sight (LOS) and non-line-of-sight (NLOS) links. The results are presented for walking, standing-sitting and running mobility conditions. In addition to transmitting power variations (from 0 dBm to −20 dBm), various configurations of the physical layers of the IEEE 802.15.6 standards are exploited. Configuration 1 (C-1) is based on 900 MHz, 101.2 Kbps and 16 bytes of packet size. Configuration 2 (C-2) and configuration 3 (C-3) differ from C-1 only by the packet sizes which are 128 bytes and 256 bytes, respectively. The last configuration (C-4) is based on

2450 MHz, 971:4 Kbps and 128 bytes of packet length. It can be seen that, at very low transmission power (such as −10 dBm and −20 dBm), the PDR starts decreasing sharply. Therefore, it is important that, while operating at very low transmission power and under realistic conditions, the packet reception performance can be significantly degraded. Consequently, such power optimizations and fine tunning have to be managed with care to ensure robust connectivity.

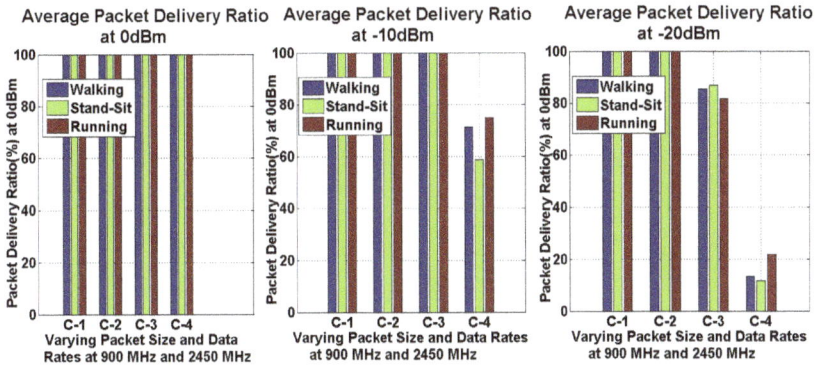

Figure 3. Average packet delivery ratio under walking, stand–sit and running mobility patterns at varied transmission power from 0 dBm, −10 dBm and −20 dBm.

3.2. Co-Channel Interference and Coexistence Techniques

With the widespread deployment of wireless networks in our daily living environment, BAN solutions are subject to strong co-channel interference, especially on the unlicensed industrial, scientific and medical (ISM) radio bands which are presently populated by various wireless technologies. The resulting interference can severely impact the connectivity and thus the communication performances.

In this regard, the IEEE 802.15.6 standard has proposed specific coexistence strategies, including beacon shifting, channel hopping, and active super-frame interleaving. In the first approach, each BAN coordinator adopts a different beacon shifting pseudo-random sequence to reduce the interference with neighboring BANs.

The second technique, which is only applicable to narrow-band channels, consists of choosing polynomial-based channel hopping sequences to avoid having neighboring BANs use the same radio channel. The active super-frame interleaving technique enables BAN coordinators to cooperatively coordinate the schedule of their active super-frames. Additionally, the carrier sense multiple access with collision avoidance (CSMA/CA) medium access control protocol could also be adopted to reduce the interference by letting the BANs nodes sense the occupancy of the radio channel prior to any data transmission.

The packet error rate distributions of co-channel interference among up-to five co-located BANs are presented in Figure 4. IEEE 802.15.6 proposed coexistence technique comparisons are highlighted. It is observed that both time shared and channel hopping approaches are well-suited to minimizing interference from neighboring BANs. However, for a dense deployment, new or enhanced schemes must be proposed.

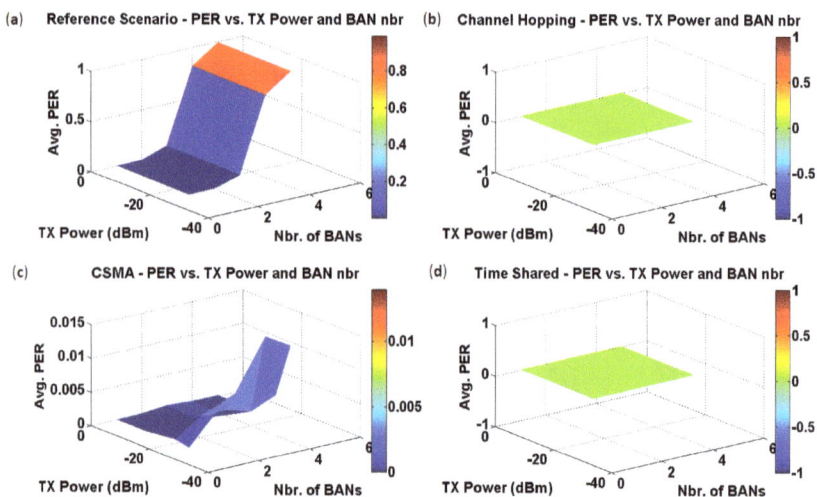

Figure 4. Packet error rate distributions of co-channel interference and evaluation of the IEEE 802.15.6 proposed coexistence techniques. (**a**) Reference scenario, without any coexistence strategy; (**b**) Channel hopping; (**c**) Carrier sense multiple access with collision avoidance (CSMA-CA)-based coexistence; (**d**) Time-shared coexistence.

3.3. Transceiver Implementations and Architecture Considerations

This section gives an overview of existing IEEE 802.15.6 transceiver (front-end and possible digital baseband part) implementations and highlights the key architectural elements typically found in such implementations. Given the high flexibility of the IEEE 802.15.6 standard, few industrial chips only implement the part of IEEE 802.15.6 standard. As illustrated below, most of the publicly documented implementations are multi-mode, e.g., supporting 802.15.6 and Bluetooth and/or Zigbee.

A few years ago, the authors of [29,30] implemented a 0.13-μm Complementary Metal-Oxide-Semiconductor (CMOS) front-end chip supporting the IEEE 802.15.6 NB PHY draft and BT-LE 4.0 standards, as well as proprietary protocols. The chip is composed of a 2.4-GHz sliding-IF receiver, a 2.4-GHz polar loop modulator transmitter, a 900-MHz loop modulator transmitter, and a low frequency 10-bit Successive AppRoximation Digital-Analog Conversion (SAR ADC) for bio-telemetry data acquisition, as well as several peripherals and digital interfaces for an Field-Programmable Gate Array (FPGA)-based digital PHY/MAC design. The chip achieves up to 1000 ksps and requires power between 1.7 and 12.3 mW, depending on the selected mode. It can operate both on the 2.36 GHz MBANs spectrum and the worldwide 2.4 GHz ISM band. It also features a transmitter for operation in China, the EU, North America and Japan (780/868/915/950 MHz, respectively).

The implementation of a 0.18 μm CMOS reconfigurable sliding-IF transceiver targeting 400 MHz/2.4 GHz IEEE 802.15.6/ZigBee WBAN hubs is presented in [31]. The receiver part comprises a wideband front-end and a reconfigurable amplifier-mixer. The transmitter part comprises a reconfigurable two stage full quadrature mixer, and a delta-sigma fractional-N Phase Locked Loop (PLL), as well as some auxiliary circuits. The chip can operate in the 0.36–0.51 GHz and the 2.36–2.5 GHz ranges. Its power consumption ranges between 13.2 mW and 18 mW.

The IEEE-802.15.6-compliant transceiver targeting a multichannel electro-acupuncture application is presented in [32]. As opposed to the two works listed above, this implementation builds upon the Human Body Communication (HBC) physical layer. The chip is implemented on a 0.13-μm CMOS process. The possible data rates are 164, 328, 626 Kb/s and 1.3125 Mb/s; its peak power is 5.5 mW (receiver-activated). The works presented in [33,34] deal with the transceiver and baseband parts, respectively, for a IEEE802.15.6/Bluetooth Low Energy/Zigbee system. The transceiver is implemented on a 90-nm CMOS process. The transmitter comprises a 2-point fractional-N PLL-based frequency modulator (FM), and a Delta-Sigma digital-controlled polar Power Amplifier (PA). The receiver builds upon a sliding-IF architecture and operates on the 2.36/2.4 GHz bands. The rates supported by the transceiver are 1 Mbps for BT-LE, 250 kbps for IEEE 802.15.4 (ZigBee), and 971 kbps for IEEE 802.15.6. It also supports a proprietary 2 Mb/s mode for data-streaming applications such as hearing aids. Its power consumption is 3.8 mW for the receiver and 4.6–4.4 mW for the transmitter.

The digital baseband part is implemented on a 40-nm low-power CMOS process. It comprises the transmitter and receiver digital baseband modules, and sub-modules responsible for processing at the PHY and Down Layer (DL) layers. Its power consumption is 200 μW for the receiver and 80 μW for the transmitter. Its data rates are identical to those of the transceiver described above. More recently, the authors of [35] designed and implemented an IEEE 802.15.6-compliant transceiver building upon the HBC physical layer (as also done in [32]). The analog front-end consists of a transmitter (transmit filter and output electrode) and a receiver (gain stage with automatic gain control and a hard decision detector). The digital part is implemented on a Xilinx Virtex 5 FPGA. The design reaches 763 Kbps (bit error rate of 0.21). The dynamic power consumption of the design is 4.5 nJ/bit with a spreading factor of 8 (it varies approximately linearly with the spreading factor).

The essential properties of these implementations are summarized in Table 2. As can be seen, implementations have been proposed for either NB or HBC PHY; typically those that support NB PHY also support BT-LE and/or Zigbee The achieved data-rates are either spot-on with the standard, or slightly below or above. However, it can be noted that the IEEE 802.15.6-compliant chipsets are yet to be commercially widely available.

Table 2. Essential properties of IEEE 802.15.6 publicly documented implementations.

Referred Works	[31]	[29,30]	[33,34]	[32]	[35]
Standard(s)	IEEE802.15.6 NB; Zigbee	IEEE802.15.6 NB; BT-LE 4.0; proprietary	IEEE802.15.6 NB; BT-LE; Zigbee	IEEE802.15.6 HBC	IEEE802.15.6 HBC
Frequency(ies)	0.36–0.51 GHz; 2.36–2.5 GHz	2.36 GHz; 2.4 GHz; 780/868/915/950 MHz	2.36 GHz; 2.4 GHz	21 MHz	21 MHz
Data-rates	N/A	1000 kbps	250/971 kbps; 2 Mb/s	164; 328; 626 Kb/s; 1.3125 Mb/s	763 kbps
Power	13.2–18 mW	1.7–12.3 mW	RX: 3.8 mW + 200 μW; TX: 4.6–4.4 mW + 80 μW	5.5 mW (peak)	4.5 nJ/bit
Front-end	0.18 μm CMOS	0.13 μm CMOS	90 nm CMOS	0.13 μm CMOS	Discrete components
Digital baseband	N/A	FPGA (not documented)	40 nm LP-CMOS	N/A	Xilinx Virtex 5

This section focuses on on-body communication. In what follows, we carry on the discussion with the next communication tier, i.e., body-to-body communication.

4. Body-to-Body Communication and Connectivity

Due to the growing number of connected devices (smart-phones, computers, game consoles, sensors, and wireless gadgets) to Internet, every human being is considered as a part of a BBN network that could be deployed at any time, anywhere in a context of Internet of humans (IoH) or Internet of things (IoT). Therefore, diverse deployment strategies are possible and various drawbacks are likely to be faced by the BBN network connectivity in a real deployment. Connectivity depends on the following factors: (1) dissemination strategy based on which data is transmitted among the wireless network; (2) communication range: which depends on the used wireless standard (i.e., BAN/MBAN,

BLE, ZigBee, etc.); (3) routing protocol which must be appropriate for the application context; and (4) experiment area: indoor/outdoor with natural/artificial obstacles. So, let us focus first on the various dissemination and routing techniques with regard to their impact on connectivity while meeting the applications requirements.

4.1. Impact of Dissemination Strategies on Connectivity

Data dissemination defines the interconnection logic hierarchy. First, we define the flat wireless network which is a set of nodes communicating with any reachable neighbors using the same wireless technology. Second, a hierarchical wireless network is a set of nodes which communicate with other neighbors from the same hierarchical level only. Specific or elected nodes only could communicate with higher level nodes. The interconnection hierarchy could be set physically (different frequencies to separate nodes into groups, different wireless technologies: more than one WiFi standard, one standard and different channels, etc.) or logically (i.e., same wireless technology and frequencies but different nodes groups related to one specific gateway node). We identify two main dissemination strategies in BBN communications related to the two different interconnection hierarchies.

The clustered data dissemination strategy consists of dividing the network into small groups of sub-networks called clusters. Each cluster leader is called a cluster head (CH). Nodes in a same cluster only communicate with their CH. CHs communicate between each others to reach wide networks. This strategy fulfills relevant operational requirements for indoor scenarios [36]. Even though this strategy defines a clear hierarchical communication charter, but it still restricts the connection capabilities only through CH. Indeed, when a CH of a cluster "A" is out of range from a neighbor cluster "B", a connection can not established even if a normal node from "A" is too close to "B". Electing new CH for the cluster "A" causes a significant delay, in addition to disconnection during the election process. Therefore, defining members for each cluster is always challenging. Consequently, connectivity of the entire BBN network depends on the connectivity between the CHs.

The distributed data dissemination strategy allows any node in the network to communicate with any reachable node. As an example, an on-body sensor placed on the right wrist could communicate with the on-body coordinator of the neighboring body. This strategy decreases the average delay of the packet compared to the clustered approach [36], whereas, it decreases the link spectral efficiency and the network overall throughput. Accordingly, the average end-to-end connectivity decreases with the link spectral efficiency.

Distributed data dissemination strategies have higher routing overhead than the clustered strategies, since in the first category, any node is allowed to send data to any node. However, in the clustered strategy, the routing overhead remains only between CHs. A delay in route refreshment in the routing table may increase routes unavailability. Therefore, a node with unavailable routes in its routing table is considered as a disconnected node.

4.2. Impact of Routing Protocols on Connectivity

In BBN networks, implemented routing protocols refer to various classes. Indeed, recent research tends to evaluate and implement ad hoc routing networks: proactive, reactive, geographic-based and gradient-based [6] in BBN communications. Moreover, context-aware protocols were proposed with regards to the implementation context (i.e, health-care, emergency, operational assistance, military, etc.). Routing protocols consist of the mechanisms that carry data from source to destination which have key role in BBN connectivity. Indeed, routing decisions are based on certain metrics depending on the use case. With regards to the study and experiments conducted by Mekikis et. al in [37], it is claimed that connectivity depends on the networking model (unicast, multicast or broadcast). Ad hoc networks, implemented in the disaster context application, use broadcast for neighbor discovery, and unicast/multicast for routing data. Based on our recent experiment [12], the network performance increases with the accurate end-to-end link quality estimations and real link conditions.

4.3. Real Indoor Experiment Scenario: A Wireless Body-to-Body Connectivity Evaluation

In order to evaluate the connectivity in indoor wireless body-to-body network scenario, we rely on the recent routing approach (i.e., Optimized Routing Approach for Critical and Emergency Networks: ORACE-Net) which is designed specifically for disaster scenario. The scenario consists of a group of people (we are considering two rescuers in this work) moving in/out an office inside a building following a disaster mobility pattern generated by Bonnmotion [38], which is a mobility scenario generation and analysis tool. Each WBAN consists of an android mobile node collecting live data from on-body Shimmer [39] sensors as depicted in Figure 5b. Additionally, four tactical static nodes (numbered from 2 to 5 in Figure 5a) are deployed during the disaster scenario which represent a temporary backbone through which data is routed to Internet. A dedicated node in the network is considered as a gateway, called a command center node (CC) placed in the back gate of the office as shown in Figure 5a. The CC relays data from the deployed network to Internet and vice versa. Our emphasis of the evaluation mainly concerns the connectivity of the WBAN mobile nodes. To that end, the tactical nodes placements are selected as such to enhance the signal propagation and as a result increase the end-to-end connectivity between mobile WBANs and Internet (through the CC node). Experiment scenario map is depicted in Figure 5a. The experimental parameters are given in Table 3.

Table 3. Experimental parameters and configuration settings. ORACE: Optimized Routing Approach for Critical and Emergency Networks; CC: command center node.

General Settings	
Parameter	**Settings**
Number of WBANs	2
ORACE-Net Tactical Devices	4 (raspberry pi 2) OS: Raspbian v8.0
Mobile nodes (coordinators)	2 (Samsung Galaxy S3-I9300-rooted) OS: Android 4.2.2 CyanogenMod 10.0
Wireless mode	Ad hoc
ESSID	CROW2
Wireless standard	IEEE 802.11n/2.412 GHz (Channel 1)
Transmission power	0 dBm
Experiment area	30 m × 150 m
CC-node connection	Ethernet to Internet Ad hoc WiFi to ORACE-Net network
Number of iterations	3
Experimentation duration	60 min/iteration
ORACE-Net Protocol and Application Layer Settings	
Application layer	MQTT client used for pushing data to the IoT platform
MQTT msg size/intervals	30 Kb/1 s
Hello/ADV packet size	20/25 Bytes
Hello/ADV intervals	3 s
Multicast address/port	224.0.0.0/10000
Shimmer [39] Sensing Device Settings	
Wireless standard	Bluetooth IEEE 802.15.1
Sensed data	Pressure, Temperature, Gyroscope $(x, y, z,$ axis-angle), Acceleration (x, y, z), Magnetometer (x, y, z), Battery level
Device/Body	1 (with multiple embedded sensors)
Buffer [39]	1024 bytes
Message interval	1 s

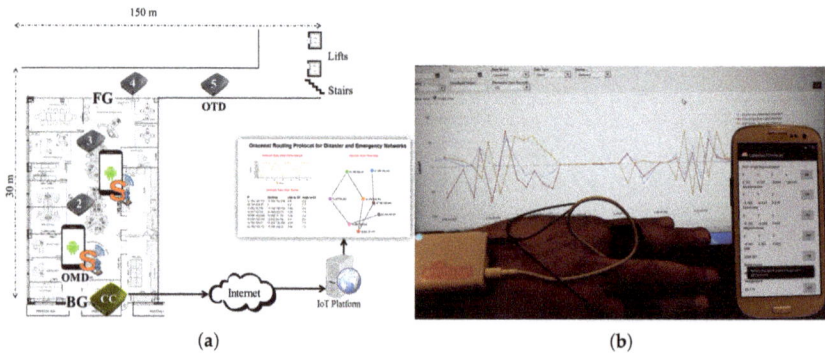

Figure 5. (a) Map of the experiment scenario; (b) Data collecting from the on-body Shimmer sensors to the android mobile application and then to the IoT platform through the ORACE-Net tactical deployed network. OMD: ORACE-Net mobile device; OTD: ORACE-Net tactical device; FG: front gate of the office; BG: back gate.

Data collected from on-body sensors is routed from WBAN node through the other WBAN node and/or tactical deployed nodes. By reaching the CC node, data is pushed through Internet to the IoT platform. On the IoT platform, data is plotted instantly as depicted by the curves shown in Figure 5b. The WBAN node behavior during the experiment is observed as depicted by Figure 6. The end-to-end link quality estimation (i.e., $E2E_{LQE}$) is a real-time metric calculated between a mobile node and the CC node. The bottom curve of Figure 6 illustrates the $E2E_{LQE}$ results over the time. There is a strong correlation between $E2E_{LQE}$ and the HOP_{Count}. It is observed that when the mobile node reaches more than 3 hops away from the CC node, and maintains that HOP_{Count} for more than 2 s, the $E2E_{LQE}$ decreases sharply. When the $E2E_{LQE}$ decreases significantly, connection latency increases and leads to mobile node disconnection. This is due to many factors: (1) signal degradation caused by the fact of being out of range (and no closed node can relay the mobile's data); and (2) the unstable links between the nodes are caused by the interference effected by WiFi access points, wireless extenders and devices inside the office. Equally important, indoor obstacles raised major signal attenuation [40]. It is noteworthy that the delay in milli-seconds (ms) depicted in Figure 6 is reset to zero when a mobile node is disconnected (we consider that a delay higher than 1000 ms is an immediate disconnection). Hence, this leads us to investigate the accuracy of the delay and disconnection times. For that, we have set up a process to ping the distant CC node every millisecond. The resulted average round-trip time delay and the average end-to-end disconnections per hop count are illustrated in blue and red respectively in Figure 7.

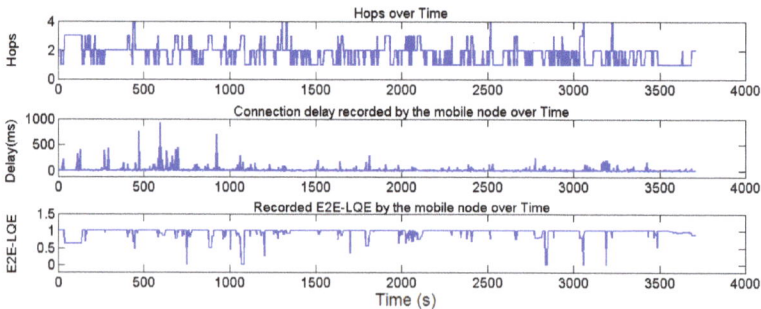

Figure 6. Hop count, instant delay and end-to-end link quality estimation variation during one hour of experimentation for WBAN node in an indoor scenario.

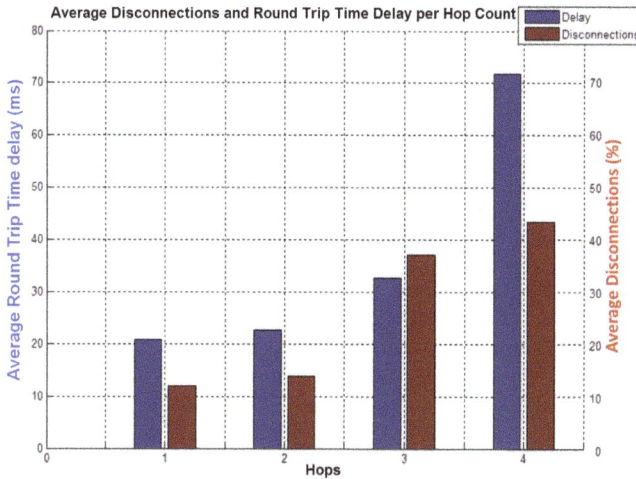

Figure 7. Average disconnections and round-trip time delay per hop count for WBAN (android smart phone mobile node with ORACE-Net protocol-enabled) in an indoor scenario.

What is important to know is that the average percentage of end-to-end disconnections and average round trip time delay increase accordingly with the hop count. With regards to the mobile smartphones used in the experiment (Samsung Galaxy S3 I9300-Battery: 2100 mAh-WiFi IEEE 802.11n), the experimental range is around 100 m. The experiment shows that the best performance is recorded within 1 hop (from the mobile node to the CC node) where average disconnection is around 12% and average round-trip time delay is equal to 21 ms. However, a connection within 4 hops (approximate distance between two nodes is 45 m) makes the average end-to-end disconnections exceed 43% as illustrated in Figure 7. The average round trip delay increases also to reach 72 ms. It is perceived that for more than 4 hops, average disconnection is expected to exceed 50%.

As has been noted, these results were achieved with intermediate tactical static nodes, hence, results might be worse if all the nodes of the network are mobile. Therefore, based on the best performance recorded (12% among 1 hop), the standard WiFi IEEE 802.11n remains efficient within short distances only.

Interestingly, we learned from our experiments that the standard WiFi IEEE 802.11n is only appropriate for body-to-body communications in the indoor scenario for an overall distance of 25 to 30 m (from one node to another). Our results and findings are consistent with the results provided by Andreev et al. in [41]. WiFi IEEE 802.11n remains a candidate for BBN communications for indoor scenarios, but with consideration of the above-cited limitation.

5. Conclusions and Perspectives

To conclude, with reference to wearable wireless network end-to-end connectivity, we highlighted the key envisioned challenges. First, while operating at low transmission power and under realistic conditions, the packet reception performance can be significantly degraded while exploiting the BAN-specific IEEE 802.15.6 standard. Second, the routing approach and the dissemination strategy have an impact on the end-to-end connectivity and the end-to-end round-trip time delay. It is concluded also, for an indoor scenario with the considered settings, that the BBN average disconnections are higher than 43% within 4 hops. Third, it is observed that wireless standard (i.e., WiFi IEEE 802.11n), while having a large coverage support, it is not entirely appropriate for BBN communications in indoor propagation as it has a very high average of end-to-end disconnections.

For the future, it is necessary to investigate other wireless body-to-body communication technologies. Indeed, WiFi IEEE 802.11ac standards, as well as cellular technologies Long Term

Evolution (LTE)/5G could potentially fulfill operational BBN requirements (especially regarding range and connectivity). Furthermore, the IEEE 802.11ac standard uses a 5-GHz frequency which avoids most of the interference possibilities and with the provision of higher number of channels defined by the standard (i.e., 25 channels by considering a channel width equal to 20 MHz) could be interesting to investigate. In addition, 4G/LTE is also another ubiquitous approach followed by 5G to deliver an edge-free body-to-body experience. Besides, specific technologies targeting the IoT could offer connectivity at various network tiers. LTE Cat M1 Machine Type Communications (eMTC), LTE CAT NB1 Narrowband Internet of Things(NB-IoT), and 802.11ah (WiFi HaLow) are such candidate technologies for which chipsets are known to exist or have been announced. Table 4 highlights a selection of such chipsets. Their availability or future availability paves the way for further investigations and modeling related to their longer range connectivity, reliability, etc. Most of these standards are designed to consume much less power than classical cellular technologies and therefore could prove very effective in future wearable wireless networks.

Table 4. Selected chipsets for upcoming future technologies.

Chipset	Ublox SARA-N2 series	Sequans Communication Monarch SQN3330	Gemalto EMS31	Intel XMM 7115/XMM 7315	Qualcomm DM9206	Newarcom NRC7191
Standard(s)	LTE Cat NB1	single-chip LTE Cat M1/NB1	LTE Cat M1	LTE CAT M1 and NB1	LTE CAT M1 and/or NB1	802.11ah (WiFi HaLow)
Data Rates	227 kbit/s DL and 21 kbit/s UL	up to 300 kbps DL/375 kbps UL in HD-FDD and 1 Mbps in FD-FDD (LTE CAT M1); up to 40 kbps DL/55 kbps UL in HD-FDD (LTE CAT NB1)	up to 300 kbps DL and 375 kbps UL	N/A	N/A	up to 2 Mbit/s

Acknowledgments: This project has received funding from the European Union's Horizon 2020 research and innovation program under grant agreement No 668995. This material reflects only the authors view and the EC Research Executive Agency is not responsible for any use that may be made of the information it contains.

Author Contributions: This paper was prepared through the collective efforts of all authors. In particular, Dhafer Ben Arbia prepared the manuscript based on the suggestions and guidance of Muhammad Mahtab Alam. Dhafer Ben Arbia made substantial contributions towards overall writing, organization, and presentation of the paper. He also conducted the experiment based on which results are discussed. In particular, he contributed in introduction, body-to-body communications: technologies, routing and challenges. He also detailed the experiment scenario and discussed the achieved results. Muhammad Mahtab Alam introduced the scope of the paper within the abstract, then presented the on-body technologies and connectivity challenges with regards to IEEE 802.15.6 standards. Yannick Le Moullec presented the different current implementations of human body communications. He also provided an overview on the future technologies with regards to the IoT trends. Elyes Ben Hamida made contributions to the literature research and paper writing. Muhammad Mahtab Alam, Yannick Le Moullec also did a critical revision of the paper and provided detailed feedback to improve the manuscript content. All authors have read and approved the final manuscript.

Conflicts of Interest: The authors declare no conflict of interest.

References

1. Alam, M.M.; Ben Hamida, E. Surveying Wearable Human Assistive Technology for Life and Safety Critical Applications: Standards, Challenges and Opportunities. *Sensors* **2014**, *14*, 9153–9209.
2. Statistics & Facts on Wearable Technology. Available online: https://www.statista.com/topics/1556/wearable-technology/ (accessed on 30 June 2017).
3. Sparks, R.; Celler, B.; Okugami, C.; Jayasena, R.; Varnfield, M. Telehealth monitoring of patients in the community. *J. Intell. Syst.* **2016**, *25*, 37–53.
4. Perlin, K. Future Reality: How Emerging Technologies Will Change Language Itself. *IEEE Comput. Gr. Appl.* **2016**, *36*, 84–89.

5. Miranda, J.; Cabral, J.; Wagner, S.R.; Fischer Pedersen, C.; Ravelo, B.; Memon, M.; Mathiesen, M. An Open Platform for Seamless Sensor Support in Healthcare for the Internet of Things. *Sensors* **2016**, *16*, 2089.

6. Ben Arbia, D.; Alam, M.M.; Attia, R.; Ben Hamida, E. ORACE-Net: A novel multi-hop body-to-body routing protocol for public safety networks. *Peer-to-Peer Netw. Appl.* **2016**, *10*, 726–749.

7. Kazemi, R.; Vesilo, R.; Dutkiewicz, E.; Fang, G. Inter-network interference mitigation in Wireless Body Area Networks using power control games. In Proceedings of the 2010 10th International Symposium on Communications and Information Technologies, Tokyo, Japan, 26–29 October 2010; pp. 81–86.

8. Smart Body Area Networks. Available online: http://www.etsi.org/technologies-clusters/technologies/smart-body-area-networks (accessed on 30 June 2017).

9. Alam, M.M.; Ben Hamida, E. Strategies for Optimal MAC Parameters Tuning in IEEE 802.15.6 Wearable Wireless Sensor Networks. *J. Med. Syst.* **2015**, *39*, 1–16.

10. Alam, M.M.; Ben Hamida, E. Wearable Wireless Sensor Networks: Applications, Standards, and Research Trends. In *Emerging Communication Technologies Based on Wireless Sensor Networks: Current Research and Future Applications*; CRC Press: Boca Raton, FL, USA, 2016; pp. 59–88.

11. Ben Arbia, D.; Alam, M.M.; Attia, R.; Ben Hamida, E. Behavior of wireless body-to-body networks routing strategies for public protection and disaster relief. In Proceedings of the IEEE 11th International Conference on Wireless and Mobile Computing, Networking and Communications (WiMob), Abu Dhabi, UAE, 19–21 October 2015; pp. 117–124.

12. Ben Arbia, D.; Alam, M.M.; Attia, R.; Ben Hamida, E. Implementation and Benchmarking of a Novel Routing Protocol for Tactical Mobile Ad hoc Networks. In Proceedings of the Third IEEE International Workshop on Emergency Networks for Public Protection and Disaster Relief (EN4PPDR 2016), New York, NY, USA, 17–19 October 2016.

13. Chen, M.; Gonzalez, S.; Vasilakos, A.; Cao, H.; Leung, V.C. Body area networks: A survey. *Mob. Netw. Appl.* **2011**, *16*, 171–193.

14. Gao, T.; Massey, T.; Selavo, L.; Crawford, D.; Chen, B.R.; Lorincz, K.; Shnayder, V.; Hauenstein, L.; Dabiri, F.; Jeng, J.; et al. The advanced health and disaster aid network: A light-weight wireless medical system for triage. *IEEE Trans. Biomed. Circuits Syst.* **2007**, *1*, 203–216.

15. Negra, R.; Jemili, I.; Belghith, A. Wireless Body Area Networks: Applications and Technologies. *Proced. Comput. Sci.* **2016**, *83*, 1274–1281.

16. Chen, X.; Xu, Y.; Liu, A. Cross Layer Design for Optimizing Transmission Reliability, Energy Efficiency, and Lifetime in Body Sensor Networks. *Sensors* **2017**, *17*, 900.

17. D'Andreagiovanni, F.; Krolikowski, J.; Pulaj, J. A fast hybrid primal heuristic for multiband robust capacitated network design with multiple time periods. *Appl. Soft Comput.* **2015**, *26*, 497–507.

18. D'Andreagiovanni, F.; Nardin, A. Towards the fast and robust optimal design of Wireless Body Area Networks. *Appl. Soft Comput.* **2015**, *37*, 971–982.

19. Aissi, H.; Bazgan, C.; Vanderpooten, D. Min–max and min–max regret versions of combinatorial optimization problems: A survey. *Eur. J. Oper. Res.* **2009**, *197*, 427–438.

20. Maniezzo, V. Exact and approximate nondeterministic tree-search procedures for the quadratic assignment problem. *INFORMS J. Comput.* **1999**, *11*, 358–369.

21. Dorigo, M.; Di Caro, G.; Gambardella, L.M. Ant algorithms for discrete optimization. *Artif. Life* **1999**, *5*, 137–172.

22. Motwani, R.; Raghavan, P. *Randomized Algorithms*; Chapman & Hall/CRC: Boca Raton, FL, USA, 2010.

23. D'Andreagiovanni, F.; Nardin, A.; Natalizio, E. A fast ILP-based Heuristic for the robust design of Body Wireless Sensor Networks. In *Applications of Evolutionary Computation*; LNCS vol. 10199; Springer: Berlin/Heidelberg, Germany, 2017; pp. 234–250.

24. D'Andreagiovanni, F.; Nace, D.; Nardin, A.; Natalizio, E. Robust relay node placement in body area networks by heuristic min-max regret. In Proceedings of the BalkanCom 2017 Conference, Tirana, Albania, 30 May–2 June 2017.

25. Tsouri, G.R.; Prieto, A.; Argade, N. On increasing network lifetime in body area networks using global routing with energy consumption balancing. *Sensors* **2012**, *12*, 13088–13108.

26. Pasquero, O.P.; Rosini, R.; D'Errico, R.; Oestges, C. A Correlation Model for Nonstationary Time-Variant On-Body Channels. *IEEE Antennas Wirel. Propag. Lett.* **2015**, *14*, 1294–1297.

27. Mani, F.; D'Errico, R. A Spatially Aware Channel Model for Body-to-Body Communications. *IEEE Trans. Antennas Propag.* **2016**, *64*, 3611–3618.
28. Alam, M.M.; Ben Hamida, E.; Ben Arbia, D.; Maman, M.; Mani, F.; Denis, B.; D'Errico, R. Realistic Simulation for Body Area and Body-To-Body Networks. *Sensors* **2016**, *16*, 561.
29. Wong, A.; Dawkins, M.; Devita, G.; Kasparidis, N.; Katsiamis, A.; King, O.; Lauria, F.; Schiff, J.; Burdett, A. A 1 V 5 mA multimode IEEE 802.15.6/bluetooth low-energy WBAN transceiver for biotelemetry applications. In Proceedings of the 2012 IEEE International Solid-State Circuits Conference, San Francisco, CA, USA, 19–23 February 2012; pp. 300–302.
30. Wong, A.C.W.; Dawkins, M.; Devita, G.; Kasparidis, N.; Katsiamis, A.; King, O.; Lauria, F.; Schiff, J.; Burdett, A.J. A 1 V 5 mA Multimode IEEE 802.15.6/Bluetooth Low-Energy WBAN Transceiver for Biotelemetry Applications. *IEEE J. Solid-State Circuits* **2013**, *48*, 186–198.
31. Zhang, L.; Jiang, H.; Wei, J.; Dong, J.; Li, F.; Li, W.; Gao, J.; Cui, J.; Chi, B.; Zhang, C.; et al. A Reconfigurable Sliding-IF Transceiver for 400 MHz/2.4 GHz IEEE 802.15.6/ZigBee WBAN Hubs with Only 21. *IEEE J. Solid-State Circuits* **2013**, *48*, 2705–2716.
32. Lee, H.; Lee, K.; Hong, S.; Song, K.; Roh, T.; Bae, J.; Yoo, H.J. A 5.5 mW IEEE-802.15.6 wireless body-area-network standard transceiver for multichannel electro-acupuncture application. In Proceedings of the 2013 IEEE International Solid-State Circuits Conference Digest of Technical Papers, San Francisco, CA, USA, 17–21 February 2013; pp. 452–453.
33. Liu, Y.H.; Huang, X.; Vidojkovic, M.; Ba, A.; Harpe, P.; Dolmans, G.; de Groot, H. A 1.9 nJ/b 2.4 GHz multistandard (Bluetooth Low Energy/Zigbee/IEEE802.15.6) transceiver for personal/body-area networks. In Proceedings of the 2013 IEEE International Solid-State Circuits Conference Digest of Technical Papers, San Francisco, CA, USA, 17–21 February 2013; pp. 446–447.
34. Bachmann, C.; van Schaik, G.J.; Busze, B.; Konijnenburg, M.; Zhang, Y.; Stuyt, J.; Ashouei, M.; Dolmans, G.; Gemmeke, T.; de Groot, H. A 0.74 V 200 µW multi-standard transceiver digital baseband in 40 nm LP-CMOS for 2.4 GHz Bluetooth Smart/ZigBee/IEEE 802.15.6 personal area networks. In Proceedings of the 2014 IEEE International Solid-State Circuits Conference Digest of Technical Papers (ISSCC), San Francisco, CA, USA, 9–13 February 2014; pp. 186–187.
35. Taylor, K. A Modular Transceiver Platform for Human Body Communications. Master's Thesis, Victoria University, Wellington, New Zealand, 2016.
36. Ben Arbia, D.; Alam, M.M.; Attia, R.; Ben Hamida, E. Data Dissemination Strategies for Emerging Wireless Body-to-Body Networks based Internet of Humans. In Proceedings of the Workshop on Advances in Body-Centric Wireless Communications and Networks and Their Applications (BCWNets 2015), Abu Dhabi, UAE, 19 October 2015.
37. Mekikis, P.V.; Kartsakli, E.; Lalos, A.S.; Antonopoulos, A.; Alonso, L.; Verikoukis, C. Connectivity of large-scale WSNs in fading environments under different routing mechanisms. In Proceedings of the 2015 IEEE International Conference on Communications (ICC), London, UK, 8–12 June 2015; pp. 6553–6558.
38. Aschenbruck, N.; Ernst, R.; Gerhards-Padilla, E.; Schwamborn, M. BonnMotion: A Mobility Scenario Generation and Analysis Tool. In Proceedings of the 3rd International Conference on Simulation Tools and Techniques, Malaga, Spain, 16–18 March 2010.
39. Burns, A.; Greene, B.R.; McGrath, M.J.; O'Shea, T.J.; Kuris, B.; Ayer, S.M.; Stroiescu, F.; Cionca, V. SHIMMER: A Wireless Sensor Platform for Noninvasive Biomedical Research. *IEEE Sens. J.* **2010**, *10*, 1527–1534.
40. Faria, D.B. Modeling signal attenuation in IEEE 802.11 wireless LANs—Vol. 1. Available online: https://pdfs.semanticscholar.org/5d18/474f224f4879a3765598713bae93f9e9c11d.pdf (accessed on 30 June 2017).
41. Andreev, S.; Galinina, O.; Pyattaev, A.; Gerasimenko, M.; Tirronen, T.; Torsner, J.; Sachs, J.; Dohler, M.; Koucheryavy, Y. Understanding the IoT connectivity landscape: A contemporary M2M radio technology roadmap. *IEEE Commun. Mag.* **2015**, *53*, 32–40.

technologies

MDPI

Article

Development of a High-Speed Current Injection and Voltage Measurement System for Electrical Impedance Tomography-Based Stretchable Sensors

Stefania Russo [1,*], Samia Nefti-Meziani [1], Nicola Carbonaro [2,3] and Alessandro Tognetti [2,3]

[1]   Autonomous System and Robotics Research Centre, University of Salford, Manchester M5 4WT, UK; s.nefti-meziani@salford.ac.uk

[2]   Research Centre E. Piaggio, University of Pisa, 56122 Pisa, Italy; nicola.carbonaro@centropiaggio.unipi.it (N.C.); a.tognetti@centropiaggio.unipi.it (A.T.)

[3]   Department of Information Engineering, University of Pisa, 56122 Pisa, Italy

*   Correspondence: s.russo1@salford.ac.uk; Tel.: +44-(0)-161-295-3231

Received: 12 July 2017; Accepted: 21 July 2017; Published: 26 July 2017

Abstract: Electrical impedance tomography (EIT) is an imaging method that can be applied over stretchable conductive-fabric materials to realize soft and wearable pressure sensors through current injections and voltage measurements at electrodes placed at the boundary of a conductive medium. In common EIT systems, the voltage data are serially measured by means of multiplexers, and are hence collected at slightly different times, which affects the real-time performance of the system. They also tend to have complicated hardware, which increases power consumption. In this paper, we present our design of a 16-electrode high-speed EIT system that simultaneously implements constant current injection and differential potential measurements. This leads to a faster, simpler-to-implement and less-noisy technique, when compared with traditional EIT approaches. Our system consists of a Howland current pump with two multiplexers for a constant DC current supply, and a data acquisition card. It guarantees a data collection rate of 78 frames/s. The results from our conductive stretchable fabric sensor show that the system successfully performs voltage data collection with a mean signal-to-noise ratio (SNR) of 55 dB, and a mean absolute deviation (MAD) of 0.5 mV. The power consumption can be brought down to 3 mW; therefore, it is suitable for battery-powered applications. Finally, pressure contacts over the sensor are properly reconstructed, thereby validating the efficiency of our EIT system for soft and stretchable sensor applications.

Keywords: EIT; stretchable; pressure sensor; conductive fabric; wearable

---

1. Introduction

Electrical impedance tomography (EIT) is a method in which an image of the internal conductivity distribution of an object is reconstructed from potential measurements made at the electrodes placed around its boundary [1,2]. In a typical procedure, a low-frequency or DC drive current is injected between two of these electrodes, and the resulting voltage data are collected from the remaining electrodes. The current injection and voltage measurements are then systematically repeated until every electrode pair has served for current injection. Once the voltage data are collected, the reconstruction of the conductivity is performed by solving the Laplacian elliptic partial differential equation [1]. Then, a finite element (FE) model of the sensor is computed, resulting in an image of the conductivity distribution.

EIT is mainly used in clinical applications for patient monitoring [3,4]; other applications include damage detection [5] and pressure sore prevention [6]. Recently, EIT has been also employed for showing the internal impedance distribution of conductive fabrics that respond to touch with local changes in conductivity, and was therefore used for developing an artificial skin as a large-area pressure sensor [7]. Such sensors have the advantage of being stretchable, and can be placed over surfaces with a different topology. In [8,9], an EIT-based sensor was placed over a mannequin arm in order to detect different types of touches. This demonstrated that these sensors have the potential of being used as wearable devices, and can now be used in robotic applications, whereby a robotic system is equipped with sensors that do not interfere with its mechanics.

However, EIT still presents a major drawback; it is considered an inverse problem, as described in [1]. Thus, EIT systems are mathematically severely ill-posed and non-linear, and are very sensitive to small changes in potential at the boundary measurements. Therefore, the image reconstruction of the internal conductivity of the body under examination is apt to errors, meaning EIT applications suffer from a low spatial resolution.

The spatial resolution can be improved by increasing the number of electrodes [10]; this creates more information available for solving the inverse problem. However, this solution affects the time required for the data collection, and therefore decreases the temporal resolution of the system.

The general approach to compensate for such a drawback is to develop data collection systems that are faster and less sensitive to noise.

Various methods have been used to increase the temporal resolution of EIT systems. In [11], the authors use a frequency-division multiplexing approach; they simultaneously inject currents at different frequencies from all the electrodes and measure the resulting voltage. In [12], a similar approach is also presented, where a fast EIT system is achieved via parallel current excitation that uses orthogonal signals. The drive currents can be then isolated, as they present diverse frequencies. Nonetheless, these approaches require synchronous analogue detection hardware and digital processing techniques that complicate the system design and increase the cost and power consumption. In [13], a fast EIT system is presented; it injects a switched DC current pulse into the drive electrode pairs and measures the voltage waveform, whereby parallel data acquisition is taken during the half part of the cycle. The problem with such a system is that, in order to achieve a fast response, the measurement time has to be small, limiting the measurement sensitivity.

A temporal resolution of 45 Hz is reported for an EIT-based sensor in [14], alongside a power consumption of roughly 22 mW. In [15], the use of a current of 10 mA at 2 kHz over a resistive material of 1 Ohm/sq results in a power loss of about 175 mW. In these approaches, the use of multiple analog switch controllers for current injection and voltage measurements complicates the hardware. Additionally, bearing in mind wearable applications, electronics should give minimum power consumption and be fit for battery-powered operation. In this paper, we present our electronic design of a printed circuit board (PCB) for 16-electrode high-speed EIT sensor applications. A video of our EIT sensor setup is available as Supplementary Materials. It works by serially injecting a constant unidirectional DC current and collecting differential voltages concurrently from all the electrodes at each current injection cycle. Parallel data acquisition consents for a higher data capture rate; therefore, this design allows for an increased temporal resolution. It also decreases the electrical common-mode noise, as the voltage data are collected in the differential mode. Another advantage of this design is that it does not have a complex hardware setup. This makes the system low-cost and of a low power consumption; thus it is more suitable for wearable applications.

The remainder of this paper is organized as follows: the electronic design of our PCB is presented in Section 2. In Section 3, we present our 16-electrode sensor system. In Section 4, we begin by showing the data acquisition frame rate along with an analysis of the voltage data; we then show different images corresponding to pressure inputs on the stretchable sensor. Finally, Section 5 concludes the paper.

2. Hardware Implementation

In EIT, a certain number of electrodes are located at the periphery of a conductive body. These electrodes serve for the application of either a small alternating or a DC current, and for performing voltage measurements; then, an image showing the internal conductivity is reconstructed using the voltage data and a FE model of the system. In Figure 1 is shown a typical EIT current injection and voltage measurement cycle. In order to scan between all electrodes and obtain a full voltage data set, the current injection and voltage measurement are systematically repeated until every electrode pair has served for the current injection. To reduce the effects of noise due to the contact impedance, voltage measurements are not performed at electrodes carrying injected current. Therefore, for a generic L-electrodes system, the number of measurements at the boundary is $K = L(L-3)$. The voltage data sets are in the form of $V^b = (V^b_1, ..., V^b_K)$, where V^b is the vector containing the voltage measurements and K is the number of measurements. For a 16-electrode system, $K = 208$. However, the total number of independent measurement is halved due to the reciprocity principle [16], as we have a symmetrical commutation of current injection and potential measurement.

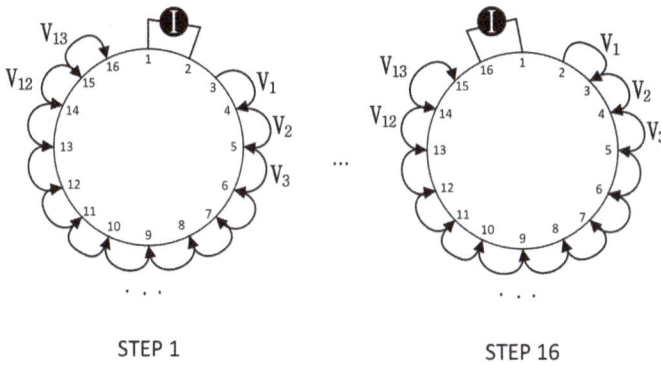

STEP 1   STEP 16

Figure 1. First and last steps of the current injection and voltage measurement sequence are shown for our 16-electrode system. For each step, the current is injected between two electrode pairs, while those remaining are used to measure the corresponding voltages. The process is repeated until every adjacent electrode pair has served for current injection.

Electronic Circuit

Our hardware system for current injection and voltage measurement is a customized PCB, as shown in Figure 2. It presents a power supply connector, a two-multiplexer mechanism connected to a Howland current pump constant current generator, a sensor block for connection with the sensor electrodes, and a connector for the National Instruments Data Acquisition (NI DAQ) USB card. The PCB can be powered with either USB or a wall block power supply.

The two multiplexers serve for the rotation of the current supply between the electrodes, and are digitally controlled by the DAQ card. They are two 16:1 ADG1606 multiplexers, presenting a typical on resistance of $4.5\,\Omega$ and a transition time, when switching from one address state to another, of $143\,\text{ns}$. With 16 channels, the multiplexers each need to be controlled by 4 digital-bit variables and 1 enabling variable, for a total of 10 variables. The first of the multiplexers has its input connected to the current source, while the latter is connected to ground; their outputs are then connected to the different electrode pairs.

The Howland current pump [17] is made of an Operational amplifier (Opamp), and provides a high-output impedance constant current source. The circuit is able to supply a constant DC current, independent of the connected load resistivity, to various loads. The Opamp for the current generator is an OP727 dual-Opamp, with a good common-mode rejection ratio (CMRR) CMRR of 85 dB, rail-to-rail, and a low supply current of $300\,\mu\text{A}/\text{amplifier}$. The DAQ card is a NI USB6353, which serves for both the multiplexer control and the parallel collection of the voltage data. The DAQ has a high input impedance, a multichannel maximum sample rate of $1.00\,\text{MS/s}$, and an ADC resolution of 16 bits. The DAQ is used for the differential voltage readings, where one input channel is connected to the positive input of the device's programmable gain instrumentation amplifier (PGIA), and the other is connected to the negative input of the PGIA; low settling times at all gains ensure the maximum resolution of the ADC is used. Differently to other works, we do not use any multiplexers for the voltage readings, but collect the data simultaneously from all the electrodes, thus decreasing electrical noise and the settling time. Furthermore, our design guarantees the possibility of switching between different types of voltage reading modalities (i.e., collecting differential voltage data between either adjacent or non adjacent electrode pairs).

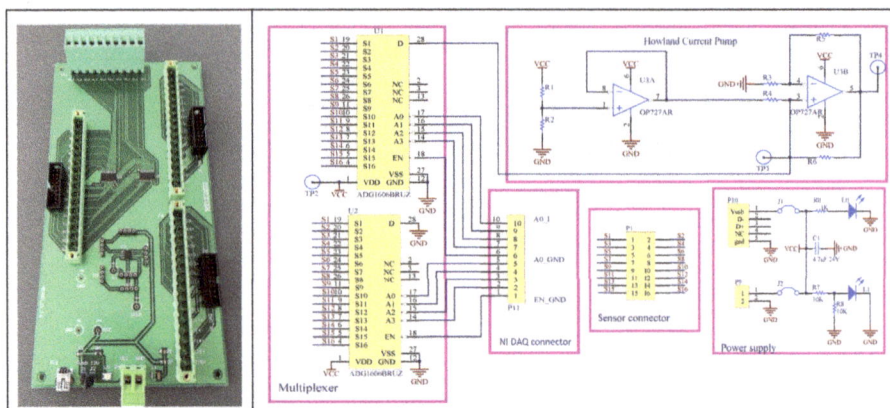

Figure 2. On the left, our customized printed circuit board (PCB) is shown. On the right, schematics of the PCB in our proposed electrical impedance tomography (EIT) system, with the sensor connection block, two multiplexers, and a Howland current pump are shown. The board can be supplied with either USB or a wall block power supply.

For the management of the DAQ card, we use the MATLAB Data Acquisition Toolbox to collect data and generate the 10 bit digital signal for the control of the two multiplexers. This approach allows for reading data into MATLAB for immediate analysis.

To further clarify the working principle of the system, a block diagram of our sensor system setup is shown in Figure 3.

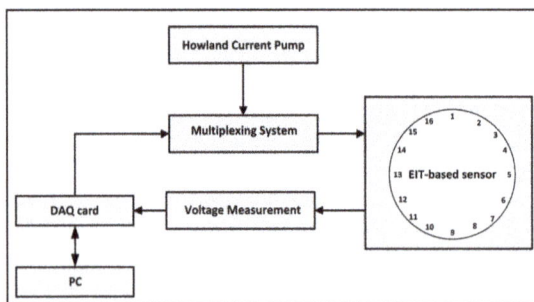

Figure 3. Block diagram of the EIT-based sensor system for current injection and voltage measurement.

3. EIT-Based Stretchable Sensor

Our EIT-based sensor is realized using a thin, stretchable piezoresistive fabric material that translates touch pressure into local changes in its resistivity. We have taken inspiration from the work of [9], and have followed our previous work [18], in which we developed the first prototype of an eight-electrode pressure-responsive sensor based on EIT.

In this paper, we present our 16-electrode sensor, which is based on a conductive stretchable fabric from the Eeonyx Corporation, made with nylon and coated with conductive doped polypyrrole (Figure 4a,b). The material has a surface resistance of 30 KΩ, and it is low-cost and lightweight. A 3D-printed circular frame made of two disc layers is used to house the conductive sheet. The frame presents 16 extrusions, where conductive copper stripes are placed to create the electrodes. The conductive fabric is then placed between the two discs, firmly in contact with the surrounding electrodes, as shown in Figure 4c.

Although a mathematical framework for an EIT-based pressure-sensor membrane was presented in [19], this model was also based on the hypothesis of incompressibility, which is not sufficient for the conductive fabric employed in this work. Here, we have relied on the model of [1], which presents a more general approach to solve the EIT inverse problem.

The image of the pressure contacts over the sensor is reconstructed by comparing two voltage data sets: V^0 is used as a background reference and V^1 is the resulting potential measured when pressure is applied. Additionally, as the DAQ card is constantly updating V^1, this method guarantees that no initial calibration is needed.

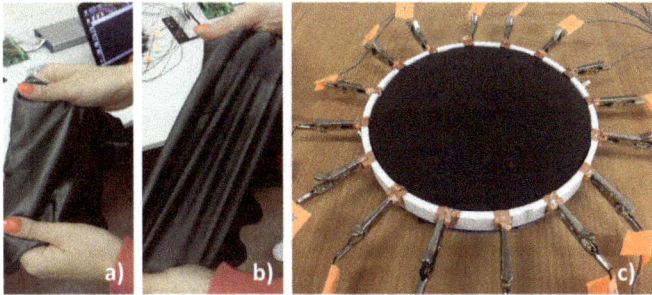

Figure 4. Our EIT-based stretchable sensor. In (**a**), the conductive fabric material in shown, and (**b**) shows the material when stretched. In (**c**), the conductive fabric is placed between the two 3D printed discs.

Image Reconstruction

The image reconstruction is carried out using a MATLAB program based on the EIDORS package [20], which is available under a General Public License. After the image ($\hat{x}_O$) of the conductivity changes is reconstructed, it requires post-processing, as it presents artefacts due to noise and possible electrode movement. To minimize these effects, we work on the image pixel values $[\hat{x}_O]_i$ and select the region of interest (ROI) in which the maximum amount of conductivity change has taken place. The processed image ($\hat{x}_P$) is found as follows:

$$[\hat{x}_P]_i = \begin{cases} [\hat{x}_O]_i & \text{if} [\hat{x}_O]_i \geq f \cdot max(\hat{x}_O) \\ 0 & \text{otherwise} \end{cases} \tag{1}$$

where $[\hat{x}_P]_i$ are the pixel values of ($\hat{x}_P$), and f is the threshold for the selection. The ROI is therefore the region of ($\hat{x}_O$) for which the pixels of ($\hat{x}_P$) are non-zero.

Is it clear that, because the choice of f determines the size of the ROI, it has a great impact on the final image, as shown in Figure 5. A number of studies have been conducted regarding the best choice of the threshold factor [7,21], but still heuristic selection is very common. In our case, we choose $f = 0.10$, as it performed best in our experiments.

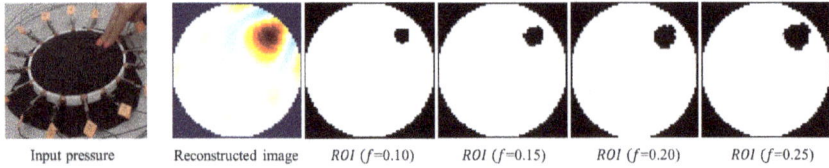

| Input pressure | Reconstructed image | *ROI (f=0.10)* | *ROI (f=0.15)* | *ROI (f=0.20)* | *ROI (f=0.25)* |

Figure 5. Examples of reconstructed images after a conductivity change has taken place on the stretchable pressure sensor. From left to right, the raw reconstructed image with a positive conductivity change (in red), and the region of interest (ROI) with different threshold factors f.

4. Measurement Results

4.1. Voltage Data Parameters

The tests shown here were conducted using the adjacent stimulation method for the current injection and measurement pattern because of its well-known use in EIT literature. In this method, two neighbours' electrodes are used for the current stimulation, while voltage data is read between the remaining adjacent electrode pairs, as seen in Figure 1. Other types of strategies can be used, but these are outside the scope of this paper. An explanation on the different drive pattern typologies can be found in [22].

The tests presented here were performed by using a current of I = 32 µA and a power supply of 16 V. This configuration gives a power consumption of about 10 mW, which is far lower than that presented in literature. This value can be further reduced to 3 mW when using a current of 10 µA supplied with a power supply of 5 V.

In order to judge the quality of the resulting signal, we use the SNR:

$$SNR = -20 \log_{10} \frac{|E[V_i]|}{\sqrt{Var(V_i)}} \tag{2}$$

where $E[V_i]$ is the mean of multiple measurements for each channel and $Var(V_i)$ is the variance between these measurements.

In Table 1, the mean SNR and the mean absolute deviation (MAD) among different measurement sets are shown for both tested currents. The results demonstrate that, in order to further reduce the power consumption, the drive current amplitude must be decreased without greatly affecting the system performance.

Table 1. Mean signal-to-noise ratio (SNR; dB) and mean absolute deviation (MAD; mV) in the case of two different current amplitudes.

	I = 32 µA	I = 10 µA
SNR	55	49
MAD	0.5	0.4

4.2. Boundary Data Collection

We chose a sample time for the voltage readings of 62,500 samples/s and selected the multiplexers' input and output channels for the 16 current injections through the digital output voltages of the DAQ. Additionally, the total number of collected samples for each channel was selected via software to be 800, which made it 50 for each injection step; this number guaranteed a data collection rate of 78 Hz, which was adequate to have a temporal resolution suitable for EIT touch-sensing applications.

Figure 6 shows an example of the voltage data set acquired through the first DAQ differential channel, namely the difference between electrodes 1 and 2 at each current injection step. The profile of the boundary data potentials indicates the effective multiplexers' channel switching with a precise conveyance of the control digital bits. Additionally, the image shows that choosing a number of samples for each channel equal to 800 is a good compromise between reaching the static conditions at each time step and having a fast data set update rate. In fact, it is visible that the effects of transients are negligible after just 25 samples. For each current injection time step, the mean value of the samples is calculated after the static conditions are reached. This is then used as the final value for calculating the 208 voltage data set, as seen in Figure 7. In order to qualitatively demonstrate the quality of the hardware setup and the voltage data, Figure 8 shows the reconstructed images when a pressure input is applied in different positions over the sensor. In the reconstructed figures, a red colour indicates a positive change in the conductivity, while a blue colour represents the ringing artefacts, which are bands or "ghosts" near the edges, typical of linear filters such as EIT systems.

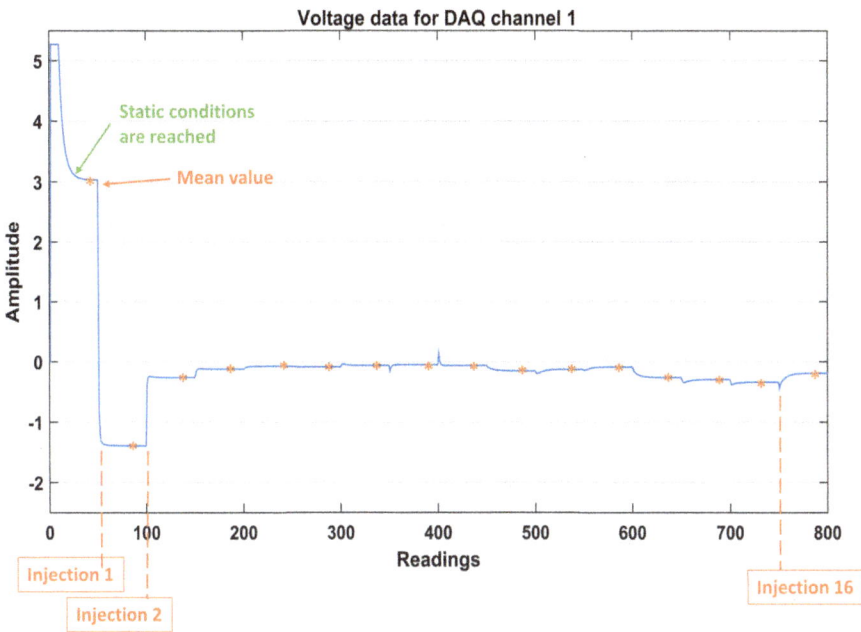

Figure 6. Measured voltage output signal on DAQ channel 1, obtained with 16 sequential DC current excitation signals. Mean values were calculated at each injection time step for each channel after the static conditions were reached, and contributed to creating the final voltage data set.

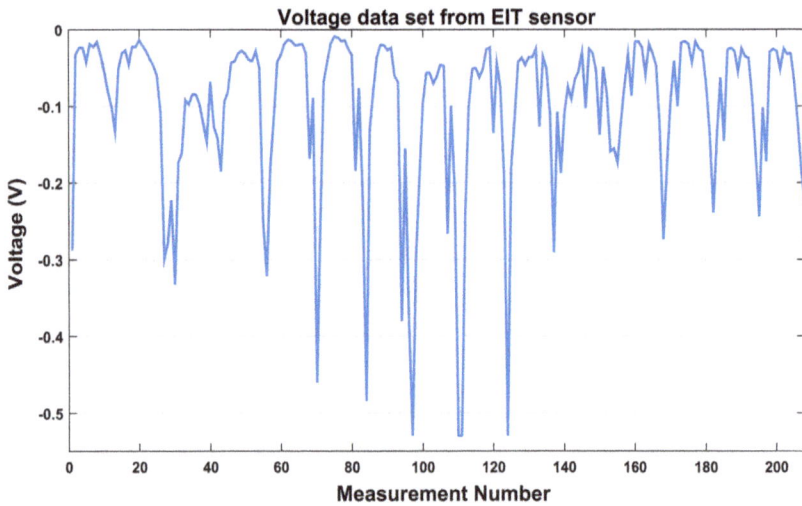

Figure 7. An example of a 208 voltage data set for a typical 16-electrode EIT system.

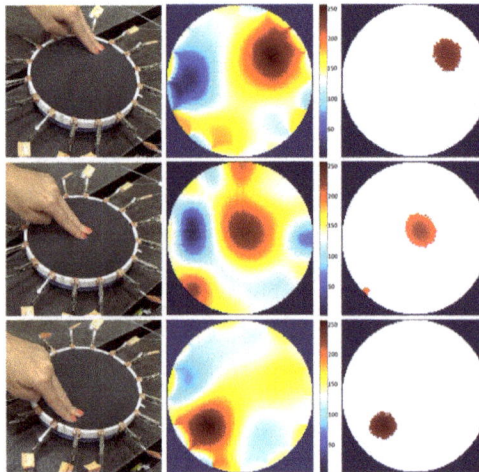

Figure 8. Reconstructed images for a pressure input applied in different locations over the sensor. On the right is the final image showing the ROI representing the maximum conductivity change. A red colour indicates an increase in the conductivity.

5. Conclusions

A high-speed EIT system that concurrently implements constant current injection and differential potential measurements has been developed for EIT-based sensor applications.

Our customized PCB design implements a Howland current pump and two analog multiplexers for the constant current injection. They are controlled using a 10 bit digital signal originating from the DAQ. The voltage data measured at the electrodes placed around the fabric sensor show the multiplexers operating in the required sequence, and also show that the data is successfully acquired for reconstructing the image of the pressure input on the stretchable sensor. The reconstructed images for different contact points over the sensor demonstrate the efficiency of our EIT-based sensor.

Our system is able to capture data at 78 frames/s and its power consumption can be brought down to 3 mW when using a current of 10 μA supplied at 5 V.

The advantages of this approach in contrast to those found in the literature are the following: (a) the less complicated hardware design makes it low-cost and more suitable for wearable systems, and (b) it has a low power consumption. Furthermore, our design is advantageous for touch-sensing applications, as its high speed guarantees that the voltage data collection is complete before any noticeable change in conductivity occurs.

Supplementary Materials: The following are available online at http://www.mdpi.com/2227-7080/5/3/48/s1: Video S1: EIT Sensor.

Acknowledgments: The research leading to these results has received funding from the People Programme (Marie Curie Actions) of the European Union Seventh Framework Programme FP7/2007-2013/ under REA grant agreement number 608022.

Author Contributions: Stefania Russo, Nicola Carbonaro and Alessandro Tognetti developed the PCB; Stefania Russo performed the experiments, analyzed the data and wrote the paper, in collaboration with Nicola Carbonaro and Alessandro Tognetti. Samia Nefti-Meziani supervised the work. All authors have read and approved the final manusript.

Conflicts of Interest: The authors declare no conflict of interest.

References

1. Holder, D.S. *Electrical Impedance Tomography: Methods, History and Applications*; Institute of Physics, IoP Publishing: Bristol, UK; Philadelphia, PA, USA, 2004.
2. Bayford, R.H. Bioimpedance tomography (electrical impedance tomography). *Annu. Rev. Biomed. Eng.* **2006**, *8*, 63–91.
3. Smith, R.W.; Freeston, I.L.; Brown, B.H. A real-time electrical impedance tomography system for clinical use-design and preliminary results. *IEEE Trans. Biomed. Eng.* **1995**, *42*, 133–140.
4. Bodenstein, M.; David, M.; Markstaller, K. Principles of electrical impedance tomography and its clinical application. *Crit. Care Med.* **2009**, *37*, 713–724.
5. Tallman, T.; Gungor, S.; Wang, K.; Bakis, C. Damage detection and conductivity evolution in carbon nanofiber epoxy via electrical impedance tomography. *Smart Mater. Struct.* **2014**, *23*, 045034.
6. Knight, R.; Lipczynski, R. The use of EIT techniques to measure interface pressure. In Proceedings of the Twelfth Annual International Conference of the IEEE, Philadelphia, PA, USA, 1–4 November 1990.
7. Silvera-Tawil, D.; Rye, D.; Soleimani, M.; Velonaki, M. Electrical impedance tomography for artificial sensitive robotic skin: A review. *IEEE Sens. J.* **2015**, *15*, 2001–2016.
8. Silvera Tawil, D.; Rye, D.; Velonaki, M. Interpretation of the modality of touch on an artificial arm covered with an EIT-based sensitive skin. *Int. J. Robot. Res.* **2012**, *31*, 1627–1641.
9. Nagakubo, A.; Alirezaei, H.; Kuniyoshi, Y. A deformable and deformation sensitive tactile distribution sensor. In Proceedings of the IEEE International Conference on Robotics and Biomimetics, Sanya, China, 15–18 December 2007.
10. Tang, M.; Wang, W.; Wheeler, J.; McCormick, M.; Dong, X. The number of electrodes and basis functions in EIT image reconstruction. *Physiol. Meas.* **2002**, *23*, 129.
11. Granot, Y.; Ivorra, A.; Rubinsky, B. Frequency-division multiplexing for electrical impedance tomography in biomedical applications. *Int. J. Biomed. Imaging* **2007**, doi:10.1155/2007/54798.
12. Gevers, M.; Gebhardt, P.; Westerdick, S.; Vogt, M.; Musch, T. Fast electrical impedance tomography based on code-division-multiplexing using orthogonal codes. *IEEE Trans. Instrum. Meas.* **2015**, *64*, 1188–1195.
13. Wilkinson, A.J.; Randall, E.; Cilliers, J.; Durrett, D.; Naidoo, T.; Long, T. A 1000-measurement frames/second ERT data capture system with real-time visualization. *IEEE Sens. J.* **2005**, *5*, 300–307.
14. Tawil, D.S. Artificial Skin and the Interpretation of Touch for Human-Robot Interaction. Ph.D. Thesis, University of Sydney, New South Wales, Australia, 2012.
15. Yao, A.; Soleimani, M. A pressure mapping imaging device based on electrical impedance tomography of conductive fabrics. *Sens. Rev.* **2012**, *32*, 310–317.
16. Geselowitz, D.B. An application of electrocardiographic lead theory to impedance plethysmography. *IEEE Trans. Biomed. Eng.* **1971**, *38* 41, doi:10.1109/TBME.1971.4502787.

17. Sheingold, D. Impedance & admittance transformations using operational amplifiers. *Lightning Empiricist* **1964**, *12*, 1–8.
18. Russo, S.; Meziani, S.N.; Gulrez, T.; Carbonaro, N.; Tognetti, A. Towards the Development of an EIT-based Stretchable Sensor for Multi-Touch Industrial Human-Computer Interaction Systems. In Proceedings of the International Conference on Cross-Cultural Design, Toronto, ON, Canada, 17–22 July 2016.
19. Ammari, H.; Kang, K.; Lee, K.; Seo, J.K. A Pressure Distribution Imaging Technique with a Conductive Membrane using Electrical Impedance Tomography. *SIAM J. Appl. Math.* **2015**, *75*, 1493–1512.
20. Adler, A.; Lionheart, W.R. Uses and abuses of EIDORS: An extensible software base for EIT. *Physiol. Meas.* **2006**, *27*, S25.
21. Naushad, A.; Rashid, A.; Mazhar, S. Analysing the performance of EIT images using the point spread function. In Proceedings of the International Conference on Emerging Technologies (ICET), Islamabad, Pakistan, 8–9 December 2014.
22. Shi, X.; Dong, X.; Shuai, W.; You, F.; Fu, F.; Liu, R. Pseudo-polar drive patterns for brain electrical impedance tomography. *Physiol. Meas.* **2006**, *27*, 1071.

MDPI

St. Alban-Anlage 66

4052 Basel

Switzerland

Tel. +41 61 683 77 34

Fax +41 61 302 89 18

www.mdpi.com

Technologies Editorial Office

E-mail: technologies@mdpi.com

www.mdpi.com/journal/technologies

www.ingramcontent.com/pod-product-compliance
Lightning Source LLC
Chambersburg PA
CBHW051904210326
41597CB00033B/6025